T0091579

Undergraduate Texts in Mathematics

Undergraduate Texts in Mathematics

Readings in Mathematics

Undergraduate Texts in Mathematics are generally aimed at third- and fourth-year undergraduate mathematics students at North American universities. These texts strive to provide students and teachers with new perspectives and novel approaches. The books include motivation that guides the reader to an appreciation of interrelations among different aspects of the subject. They feature examples that illustrate key concepts as well as exercises that strengthen understanding.

Sidney A. Morris • Arthur Jones •
Kenneth R. Pearson

Abstract Algebra and Famous Impossibilities

Squaring the Circle, Doubling the Cube, Trisecting an Angle, and Solving Quintic Equations

Second Edition

With 28 Illustrations

Sidney A. Morris
Department of Mathematical
and Physical Sciences
La Trobe University
Bundoora, VIC, Australia

School of Engineering
IT and Physical Sciences
Federation University Australia
Ballarat, VIC, Australia

Kenneth R. Pearson (Deceased)
VIC, Australia

Arthur Jones (Deceased)
VIC, Australia

ISSN 0172-6056 ISSN 2197-5604 (electronic)
Undergraduate Texts in Mathematics
ISSN 2945-5839 ISSN 2945-5847 (electronic)
Readings in Mathematics
ISBN 978-3-031-05700-7 ISBN 978-3-031-05698-7 (eBook)
https://doi.org/10.1007/978-3-031-05698-7

Mathematics Subject Classification: 12-01, 12F05, 12E05, 01A55

The first edition of this textbook was published in the series Universitext.

This Springer imprint is published by the registered company Springer Nature Switzerland AG
The registered company address is: Gewerbestrasse 11, 6330 Cham, Switzerland

Preface to the Second Edition

It is thirty years since the first edition of this book appeared. A distinguishing feature of the first edition and this edition is that the study of abstract algebra is motivated by demonstrating that it is exactly what was needed to solve several famous problems which had remained unsolved for over 2,000 years.

In the first edition we showed how the three geometric problems known as Doubling the Cube, Trisecting an Angle, and Squaring the Circle were finally solved. In this edition, a famous fourth ancient problem, namely Solving Polynomial Equations, is added to this list.

In the last chapter of the first edition we touched briefly on how the one thousand year old calculus problem of Integration in Closed Form which dates back to Newton and Leibniz, the founders of calculus, was solved using abstract algebra. In this edition the explanation is significantly expanded by introducing the key concept of differential field.

That π and e are transcendental numbers was proved in the first edition. In this edition exciting advances in transcendental number theory which have occurred in the 130 years since Lindemann proved that π is transcendental are included. I thank Taboka Prince Chalebgwa for advice on the section on Transcendental Number Theory.

As students were very complimentary about the gentle style of presentation in the first edition, every attempt has been made to maintain that in this edition. Another feature of this edition is the inclusion of much more historical information. The bibliography has also been greatly expanded.

In several places in the book there are Proofs by Contradiction. To highlight when this type of proof is being used the word "suppose" is reserved for the beginning of a proof by contradiction. In other proofs the word "assume" appears rather than "suppose".

The first edition grew out of a second-year subject which Arthur Jones, Ken Pearson, and I had taught for many years at La Trobe University in Melbourne, Australia. This second edition is aimed at similar students. But it should also be mentioned that the knowledge in this book would be beneficial to future (and current) high school teachers. The level of the material also means that it is accessible to talented high school students.

The authors of the first edition were Arthur Jones, Ken Pearson, and me. Arthur and Ken died in 2006 and 2015, respectively. While writing this second edition, I missed them greatly—each had a great passion for teaching and for mathematics.

In preparing this edition I was assisted by the advice from the Editors of this Series. I am also grateful to Springer whose staff provided a LaTeX version of the first edition as a starting point for the second edition. I am extremely grateful to the Springer Mathematics Editor, Dr Loretta Bartolini, who displayed much professionalism and patience.

Last but not least, I thank my wife Elizabeth for her continued support over four decades of marriage.

Bundoora/Ballarat, Australia
March 2022

Sidney A. Morris

Preface to the First Edition

The famous problems of squaring the circle, doubling the cube, and trisecting an angle captured the imagination of both professional and amateur mathematicians for over two thousand years. Despite the enormous effort and ingenious attempts by these men and women, the problems would not yield to purely geometrical methods. It was only the development of abstract algebra in the nineteenth century which enabled mathematicians to arrive at the surprising conclusion that these constructions are not possible.

In this book we develop enough abstract algebra to prove that these constructions are impossible. Our approach introduces all the relevant concepts about fields in a way which is more concrete than usual and which avoids the use of quotient structures (and even of the Euclidean algorithm for finding the greatest common divisor of two polynomials). Having the geometrical questions as a specific goal provides motivation for the introduction of the algebraic concepts and we have found that students respond very favourably.

We have used this text to teach second-year students at La Trobe University over a period of many years, each time refining the material in the light of student performance.

The text is pitched at a level suitable for students who have already taken a course in linear algebra, including the ideas of a vector space over a field, linear independence, basis and dimension. The treatment, in such a course, of fields and vector spaces as algebraic objects should provide an adequate background for the study of this book. Hence the book is suitable for Junior/Senior courses in North America and second-year courses in Australia.

Chapters 1 to 6, which develop the link between geometry and algebra, are the core of this book. These chapters contain a complete solution to the three famous problems, except for proving that π is a transcendental number (which is needed to complete the proof of the impossibility of squaring the circle). In Chapter 7 (Chapter 10 in the second edition) we give a self-contained proof that π is transcendental. Chapter 8 (Chapter 11 in the second edition) contains material about fields which is closely related to the topics in Chapters 2–4, although it is not required in the proof of the impossibility of the three constructions. The short concluding Chapter 9 (Chapter 12 in the second edition) describes some other areas of mathematics in which algebraic machinery can be used to prove impossibilities.

We expect that any course based on this book will include all of Chapters 1–6 and (ideally) at least passing reference to Chapter 9 (Chapter 12 in the second edition). We have often taught such a course which we cover in a term (about twenty hours). We find it essential for the course to be paced in a way that allows time for students to do a substantial number of problems for themselves. Different semester length (or longer) courses including topics from Chapters 7 and 8 (Chapters 11 and 12 in the second edition) are possible. The three natural parts of these are

(1) Sections 7.1 and 7.2 (Sections 10.1 and 10.2 in the second edition) (transcendence of e),
(2) Sections 7.3 to 7.6 (10.3 to 10.6.14 in the second edition) (transcendence of π),
(3) Chapter 8 (Chapter 11 in the second edition).

These are independent except, of course, that (2) depends on (1). Possible extensions to the basic course are to include one, two or all of these. While most treatments of the transcendence of π require familiarity with the theory of functions of a complex variable and complex integrals, ours in Chapter 7 (Chapter 10 in the second edition) is accessible to students who have completed the usual introductory real calculus course (first-year in Australia and Freshman/Sophomore in North America). However instructors should note that the arguments in Sections 7.3 to 7.6 (Sections 10.3 to 10.6.14 in the second edition) are more difficult and demanding than those in the rest of the book.

Problems are given at the end of each section (rather than collected at the end of the chapter). Some of these are computational and others require students to give simple proofs.

Each chapter contains additional reading suitable for students and instructors. We hope that the text itself will encourage students to do further reading on some of the topics covered.

As in many books, exercises marked with an asterisk * are a good bit harder than the others. We believe it is important to identify clearly the end of each proof and we use the symbol ■ for this purpose.

We have found that students often lack the mathematical maturity required to write or understand simple proofs. It helps if students write down where the proof is heading, what they have to prove and how they might be able to prove it. Because this is not part of the formal proof, we indicate this exploration by separating it from the proof proper by using a box which looks like

> Assume we are asked to prove that some result is true for every positive integer n. We might first look at some special cases such as $n = 1$, $n = 2$, $n = 3$. We may see that the result is true in these three cases. Of course this does not mean it is true for all positive integers n. While what we have written may not form part of a formal proof, it may nevertheless give us a hint to how to proceed to prove the general result.

Experience has shown that it helps students to use this material if important theorems are given specific names which suggest their content. We have enclosed these names in square brackets before the statement of the theorem. We encourage students to use these names when justifying their solutions to exercises. They often find it convenient to abbreviate the names to just the relevant initials. (For example, the name "Small Degree Irreducibility Theorem" can be abbreviated to S.D.I.T.)

We are especially grateful to our colleague Gary Davis, who pointed the way towards a more concrete treatment of field extensions (using residue rings rather than quotient rings) and thus made the course accessible to a wider class of students. We are grateful to Ernie Bowen, Jeff Brooks, Grant Cairns, Mike Canfell, Brian Davey, Alistair Gray, Paul Halmos, Peter Hodge, Alwyn Horadam, Deborah King, Margaret McIntyre, Bernhard H. Neumann, Kristen Pearson, Suzanne Pearson, Alf van der Poorten, Brailey Sims, Ed Smith and Peter Stacey, who have given us helpful feedback, made suggestions and assisted with the proof reading.

We thank Dorothy Berridge, Ernie Bowen, Helen Cook, Margaret McDonald and Judy Storey for skilful TeXing of the text and diagrams, and Norman Gaywood for assisting with the index.

April 1991

Arthur Jones
Sidney A. Morris
Kenneth R. Pearson

Introduction

0.1 Four Famous Problems

In this book we discuss four of the oldest problems in mathematics. Each of them is over 2,000 years old. The four problems are known as:

[I] doubling the cube (or duplicating the cube, or the Delian problem);
[II] trisecting an arbitrary angle;
[III] squaring the circle (or quadrature of the circle);
[IV] solving polynomial equations.

Problem I is to construct a cube having twice the volume of a given cube. Problem II is to describe how every angle can be trisected. Problem III is that of constructing a square whose area is equal to that of a given circle. In each of these cases, the constructions are to be carried out using only a ruler and compass.

Reference to Problem I occurs in the following ancient document supposedly written by Eratosthenes to King Ptolemy III about the year 240 B.C.E.:

To King Ptolemy, Eratosthenes sends greetings. It is said that one of the ancient tragic poets represented Minos as preparing a tomb and as declaring, when he learnt it was a hundred feet each way: "Small indeed is the tomb thou hast chosen for a royal burial. Let it be double [in volume]. And thou shalt not miss that fair form if thou quickly doublest each side of the tomb". But he was wrong. For when the sides are doubled, the surface [area] becomes four times as great, and the volume eight times. It became a subject of inquiry amongst geometers in what manner one might double the given volume without changing the shape. And this problem was called the duplication of the cube, for given a cube they sought to double it...

The origins of Problem II are obscure. The Greeks were concerned with the problem of constructing regular polygons, and it is likely that the trisection problem arose in this context. This is so because the construction of a regular polygon with nine sides necessitates the trisection of an angle.

The history of Problem III is linked to that of calculating the area of a circle. Information about this is contained in the Rhind Papyrus, perhaps the best known ancient mathematical manuscript, which was brought by the Scottish archaeologist Alexander Henry Rhind (1833–1863) to the British Museum in the nineteenth century. The manuscript was copied by the Egyptian scribe Ahmes about 1650 B.C.E. from an even older work. It states that the area of a circle is equal to that of a square whose side is the diameter diminished by one ninth; that is, $A = \left(\frac{8}{9}\right)^2 d^2$. Comparing this with the formula $A = \pi r^2 = \pi \frac{d^2}{4}$ gives

$$\pi = 4 \cdot \left(\frac{8}{9}\right)^2 = \frac{256}{81} = 3.1604\ldots$$

The Papyrus contains no explanation of how this formula was obtained. Fifteen hundred years later Archimedes, a Greek mathematician from Syracuse, Italy, showed that

$$3\frac{10}{71} < \pi < 3\frac{10}{70} = 3\frac{1}{7}.$$

(Note that $3\frac{10}{71} = 3.14084\ldots, 3\frac{1}{7} = 3.14285\ldots, \pi = 3.14159\ldots$.)

Throughout the ages these problems were tackled by most of the best mathematicians. For some reason, amateur mathematicians were also fascinated by them. In the time of the Greeks a special word was used to describe people who tried to solve Problem III—$\tau\varepsilon\tau\rho\alpha\gamma\omega\nu\iota\zeta\varepsilon\iota\nu$ (tetragonidzein) which means *to occupy oneself with the quadrature*.

In 1775 the Paris Academy found it necessary to protect its officials from wasting their time and energy examining purported solutions of these problems by amateur mathematicians. It passed a resolution (Histoire de l'Académie royale, année 1775, p. 61) that no more solutions were to be examined of the problems of doubling the cube, trisecting an arbitrary angle, and squaring the circle and that the same resolution should apply to machines for exhibiting perpetual motion. (See (Hudson, 1953, p. 3).)

The problems were finally solved in the nineteenth century. In 1837, the French mathematician Pierre Wantzel (1814–1848) settled Problems I and II. In 1882, the German mathematician Ferdinand von Lindemann (1852–1939) disposed of Problem III.

0.2 Straightedge and Compass Constructions

Construction problems are, and have always been, a favourite topic in geometry. Using only a ruler and compass, a great variety of constructions is possible. Some of these constructions are described in detail in Section 5.1:

a line segment can be bisected; any angle can be bisected; a line can be drawn from a given point perpendicular to a given line; etc.

In all of these problems *the ruler is used merely as a straightedge, an instrument for drawing a straight line but not for measuring or marking off distances*. Such a ruler will be referred to as a *straightedge*.

Problem I is that of constructing with compass and straightedge a cube having twice the volume of a given cube. If the side of the given cube has length 1 unit, then the volume of the given cube is $1^3 = 1$. So the volume of the larger cube should be 2, and its sides should thus have length $\sqrt[3]{2}$. *Hence the problem is reduced to that of constructing, from a segment of length 1, a segment of length* $\sqrt[3]{2}$.

Problem II is to produce a construction for trisecting any given angle. While it is easy to give examples of particular angles which can be trisected, the problem is to give a construction which will work for **every** angle.

Problem III is that of constructing with compass and straightedge a square of area equal to that of a given circle. If the radius of the circle is taken as one unit, the area of the circle is π, and therefore the area of the constructed square should be π; that is, the side of the square should be $\sqrt{\pi}$. *So the problem is reduced to that of constructing, from a segment of length 1, a segment of length* $\sqrt{\pi}$.

0.3 Impossibility of the Geometric Constructions

Why did it take so many centuries for these problems to be solved? The reasons are (i) the required constructions are **impossible**, and (ii) a full understanding of these problems comes not from geometry but from abstract algebra (a subject not born until the nineteenth century). Our purpose is to introduce this algebra and show how it is used to prove the impossibility of these constructions.

A real number γ is said to be *constructible* if, starting from a line segment of length 1, we can construct a line segment of length $|\gamma|$ in a finite number of steps using straightedge and compass.

We shall prove, in Chapters 5 and 6, that *a real number is constructible if and only if it can be obtained from the number 1 by successive applications of the operations of addition, subtraction, multiplication, division, and taking square roots*. Thus, for example, the number

$$2+\sqrt{3+10\sqrt{2}}$$

is constructible.

Now $\sqrt[3]{2}$ does not appear to have this form. Appearances can be deceiving, however. How can we be sure? The answer turns out to be that if $\sqrt[3]{2}$ did have this form, then a certain vector space would have the wrong dimension! This settles Problem I.

As for Problem II, note that it is sufficient to give just one example of an angle which cannot be trisected. One such example is the angle of $60°$. It can be shown that this angle can be trisected only if $\cos 20°$ is a constructible number. But, as we shall see in Chapter 6, the number $\cos 20°$ is a solution of the cubic equation

$$8x^3 - 6x - 1 = 0$$

which does not factorise over the rational numbers. Hence it seems likely that *cube roots*, rather than square roots, will be involved in its solution, so we would not expect $\cos 20°$ to be constructible. Once again this can be made into a rigorous proof by considering the possible dimensions of a certain vector space.

As we shall show in Chapter 6, the solution of Problem III also hinges on the dimension of a vector space. Indeed, the impossibility of squaring the circle follows from the fact that a certain vector space (a different one from those mentioned above in connection with Problems I and II) is not finite-dimensional. This in turn is because the number π is “transcendental”, which we shall prove in Chapter 10.

0.4 Solving Polynomial Equations

Problem IV is that of solving polynomial equations of degree 2, 3, 4, and, in particular, 5 and above. Thousands of years ago the Babylonians, Chinese, Egyptians, and Greeks knew how to find *some* solutions of *some* quadratic equations. Cubic equations had also been examined by the Babylonians, Chinese, Egyptians, Greeks, and Indians and *some* solutions of *particular* cubic equations were known to them. However, it was not until the year 628 that an explicit, but not completely general, solution of $ax^2 + bx = c$, was produced by the Indian mathematician Brahmagupta (590–668). The complete solution of quadratic equations was published in 1594 by the Flemish mathematician Simon Stevin (1548–1620). In the sixteenth century the Italian mathematician Scipione del Ferro (1465–1526) discovered how to solve a wide class of cubic equations. About 1540 the Italian mathematician Lodovico Ferrari (1522–1565) discovered how to solve quartic equations. The combined work of the Italian mathematician Paolo Ruffini (1765–1822) in 1798 and the Norwegian mathematician Niels Henrik Abel (1802–1829) in 1826 proved that there are polynomial equations of degree 5 and above which are impossible to solve (in terms of radicals). These are discussed in Chapters 7–9.

Additional Reading for the Introduction

More information about the background to the three geometric problems can be found in various books on the history of mathematics including Bell (1937, 1945); Kline (1972); Struik (1967); Sanford (1958).

References dealing specifically with Greek mathematics include Gow (1968), Heath (1921) and Lasserre (1964). A detailed history of Problem III is given in Hobson (1953).

The original solutions to Problems I and II by Wantzel are in Wantzel (1837) and the original solution to Problem III by Lindemann is in Lindemann (1882).

The book Stewart (2004) has a Historical Introduction dealing with Problem IV.

Contents

List of Figures

Chapter 1
Algebraic Preliminaries

This chapter presents the background algebra on which the rest of this book depends. Much of this material should be familiar to you.

The Rational Roots Test, however, will probably be new to you. It provides a nice illustration of how polynomials can be used to study certain properties of real numbers; for example, we use it to show that the numbers $\sqrt[5]{2}$, $\sqrt{2}+\sqrt{3}$, and $\sin 20^\circ$ are irrational. The application of polynomials to the study of numbers will be an important theme throughout the book.

1.1 Fields, Rings and Vector Spaces

In this section we summarize the main ideas and terminology of fields, rings, and vector spaces which we shall use throughout this book. If you are not already familiar with some of this material, we refer you to the Appendix to Chapter 1 for more details.

Fields

A familiar example of a *field* is the set $\mathbb{Q}$ of all rational numbers. For this set, the usual operations of arithmetic

addition, subtraction, multiplication, and division (except by 0)

can be performed without restriction. The same is true of the set $\mathbb{R}$ of all real numbers, which is another example of a field. Likewise the set $\mathbb{C}$ of all complex numbers, with the above operations, is a field.

A formal definition of a field is given in the Appendix to Chapter 1. Strictly speaking, a field consists of a set $\mathbb{F}$ together with the operations to be performed on $\mathbb{F}$.

S. A. Morris et al., *Abstract Algebra and Famous Impossibilities*, Undergraduate Texts in Mathematics, https://doi.org/10.1007/978-3-031-05698-7_1

When there is no ambiguity as to which operations are intended, we simply refer to the set $\mathbb{F}$ as the field.

If $\mathbb{F}$ is a field and $\mathbb{E} \subseteq \mathbb{F}$ it may happen that $\mathbb{E}$ also becomes a field when we apply the operations on $\mathbb{F}$ to its elements. If this happens we say that $\mathbb{E}$ is a *subfield* of $\mathbb{F}$ and, reciprocally, that $\mathbb{F}$ is an *extension field* of $\mathbb{E}$. This situation is often denoted by $\mathbb{F}/\mathbb{E}$ and read as "$\mathbb{F}$ over $\mathbb{E}$". As particular cases, $\mathbb{Q}$ is a subfield of both $\mathbb{C}$ and $\mathbb{R}$ while $\mathbb{R}$ and $\mathbb{C}$ are both extension fields of $\mathbb{Q}$. We write this as $\mathbb{C}/\mathbb{Q}$ and $\mathbb{R}/\mathbb{Q}$.

An important relationship between $\mathbb{Q}$ and $\mathbb{C}$ is expressed in the following proposition.

1.1.1 Proposition. *The field $\mathbb{Q}$ of all rational numbers is the smallest subfield of $\mathbb{C}$, the field of all complex numbers.*

Proof. We are to show that if $\mathbb{F}$ is a subfield of $\mathbb{C}$ then $\mathbb{Q} \subseteq \mathbb{F}$. So we let $\mathbb{F}$ be any subfield of $\mathbb{C}$.

To prove that $\mathbb{Q} \subseteq \mathbb{F}$ we shall show that

$$\text{if} \quad x \in \mathbb{Q} \quad \text{then} \quad x \in \mathbb{F}.$$

Let $x \in \mathbb{Q}$; that is,

$$x = \frac{r}{s}, \quad \text{where } r, s \text{ are integers and } s \neq 0.$$

Our aim is to prove that $x \in \mathbb{F}$.

To do this we use the field properties of $\mathbb{F}$. In particular, every field is closed under addition, subtraction, multiplication, and division.

As $\mathbb{F}$ is a field, the number 1 must be in $\mathbb{F}$.
Therefore each positive integer is in $\mathbb{F}$, as $\mathbb{F}$ is closed under addition.
So each negative integer is in $\mathbb{F}$, as $\mathbb{F}$ is closed under subtraction.
Also $0 \in \mathbb{F}$, as $\mathbb{F}$ is a field.
Hence the integers r and s are in $\mathbb{F}$.
Since $\mathbb{F}$ is closed under division (and $s \neq 0$), the quotient $r/s \in \mathbb{F}$.
So $x \in F$, as required. ■

There are, in fact, lots of interesting fields which lie between $\mathbb{Q}$ and $\mathbb{C}$ and we shall devote much time to studying them.

Of course not all fields are subfields of $\mathbb{C}$. To see this, consider the following. For each positive integer n put

$$\mathbb{Z}_n = \{0, 1, 2, \ldots, n-1\}.$$

We define addition of elements a and b in $\mathbb{Z}_n$ by

$$a \oplus_n b = a + b \pmod{n};$$

that is, add a and b in the usual way and then subtract multiples of n until the answer lies in the set $\mathbb{Z}_n$. We define multiplication similarly:

$$a \otimes_n b = a \times b \pmod{n}.$$

For example,

$$\begin{aligned} 3 \oplus_6 4 &= 1 \quad \text{(subtract 6 from 7)},\\ 2 \oplus_6 1 &= 3,\\ 3 \otimes_6 4 &= 0 \quad \text{(subtract 12 from 12)},\\ 9 \otimes_{10} 7 &= 3. \end{aligned}$$

It can be shown that if n is a prime number then $\mathbb{Z}_n$ is a field. It is obvious that in this case $\mathbb{Z}_n$ is not a subfield of $\mathbb{C}$. For example, if $n = 2$, $1 \oplus_2 1 = 0$ in $\mathbb{Z}_2$, but $1 + 1 = 2$ in $\mathbb{C}$; so the operation of addition in the field $\mathbb{Z}_2$ is different from that in the field $\mathbb{C}$. Unlike our previous examples $\mathbb{Q}$, $\mathbb{R}$ and $\mathbb{C}$, the field $\mathbb{Z}_n$ is a *finite field*; that is, it has only a finite number of elements.

Rings

The concept of a ring is less restrictive than that of a field in that we no longer require the possibility of division but only of

addition, subtraction, and multiplication.

Thus the set $\mathbb{Z}$ of all the integers is an example of a ring which is not a field, whereas the set $\mathbb{N}$ of all natural numbers is not a ring.

We even allow the concept of multiplication itself to be liberalized. We do not require it to be commutative; that is, ab may be unequal to ba. We also permit zero divisors, so that it is possible to have

$$ab = 0 \quad \textit{but neither } a \textit{ nor } b \textit{ is zero}.$$

Observe that if M is the set of all 2×2 matrices with entries from $\mathbb{R}$, then M is a ring with the ring operations being matrix addition and matrix multiplication. However, M is not a field since if

$$A = \begin{pmatrix} 1 & 2 \\ 3 & 4 \end{pmatrix} \quad \text{and} \quad B = \begin{pmatrix} 5 & 6 \\ 7 & 8 \end{pmatrix}$$

$$\text{then} \quad AB = \begin{pmatrix} 19 & 22 \\ 43 & 50 \end{pmatrix} \neq BA = \begin{pmatrix} 23 & 34 \\ 31 & 46 \end{pmatrix}.$$

Note also that this ring M has zero divisors; for example, if

$$C = \begin{pmatrix} 1 & 0 \\ 0 & 0 \end{pmatrix} \quad \text{and} \quad D = \begin{pmatrix} 0 & 0 \\ 4 & 5 \end{pmatrix} \quad \text{then} \quad CD = \begin{pmatrix} 0 & 0 \\ 0 & 0 \end{pmatrix}.$$

A formal definition of a ring is given in the Appendix to Chapter 1.

Vector Spaces

Recall from linear algebra that a *vector space* consists of a set V together with a field $\mathbb{F}$. The elements of V are called *vectors* and the elements of $\mathbb{F}$ are called *scalars*. Any two vectors can be added to give a vector and a vector can be multiplied by a scalar to give a vector. A formal definition of a vector space is given in the Appendix to Chapter 1.

When there is no danger of ambiguity it is customary to omit reference to both the scalars and the operations and to regard V itself as the vector space. In our study, however, the field of scalars will be constantly changing and to omit reference to it would almost certainly lead to ambiguity. Hence we shall include the field $\mathbb{F}$ in our description and refer to "*the vector space V over $\mathbb{F}$*".

Two other notions which you need to recall are those of spanning and linear independence.

If $v_1, v_2, \ldots, v_n$ are vectors in the vector space V over the field $\mathbb{F}$, then we say that the set $\{v_1, v_2, \ldots, v_n\}$ of these vectors *spans V over $\mathbb{F}$* providing that every vector $v \in V$ can be written as a linear combination of $v_1, v_2, \ldots, v_n$; that is,

$$v = \lambda_1 v_1 + \lambda_2 v_2 + \ldots + \lambda_n v_n$$

for some $\lambda_1, \lambda_2, \ldots, \lambda_n$ in $\mathbb{F}$.

A vector space is said to be *finite-dimensional* if there is a finite set of vectors $\{v_1, v_2, \ldots, v_n\}$ which spans it.

The set $\{v_1, v_2, \ldots, v_n\}$ of vectors is said to be *linearly independent over* $\mathbb{F}$ if the zero vector, 0, can be written as a linear combination of $v_1, v_2, \ldots, v_n$ in only one way, namely with all the coefficients equal to zero; that is,

$$0 = \lambda_1 v_1 + \lambda_2 v_2 + \ldots + \lambda_n v_n \quad (\lambda_i \in \mathbb{F}) \implies \lambda_1 = \lambda_2 = \ldots = \lambda_n = 0.$$

The set $\{v_1, v_2, \ldots, v_n\}$ is said to be a *basis* for V over $\mathbb{F}$ if it is linearly independent over $\mathbb{F}$ and spans V over $\mathbb{F}$.

While a finite-dimensional vector space can have an infinite number of distinct bases, the number of vectors in each of the bases is the same and is called the *dimension* of the vector space. Recall also that, in a vector space of dimension n, any set with more than n vectors must be linearly dependent. (For example, any set of four vectors in a three-dimensional space is linearly dependent.)

Pairs of Fields

The way in which vector spaces enter our study is as pairs of fields, one of which is a subfield of the other.

The most intuitive example, in this regard, is the vector space $\mathbb{C}$ over $\mathbb{R}$ in which we take $\mathbb{C}$ for the vectors and $\mathbb{R}$ for the scalars. This vector space arises from the pair of fields $\mathbb{C}$ and $\mathbb{R}$, in which $\mathbb{R}$ is a subfield of $\mathbb{C}$. We may think of the elements of $\mathbb{C}$ as points in the plane and then the vector space operations correspond exactly to the usual vector addition and scalar multiplication in the plane.

More generally, if $\mathbb{E}$ and $\mathbb{F}$ are fields with $\mathbb{E}$ a subfield of $\mathbb{F}$, we may take $\mathbb{F}$ for the vectors and $\mathbb{E}$ for the scalars to get the *vector space* $\mathbb{F}$ *over* $\mathbb{E}$. (Note that the smaller field is the scalars and the larger one is the vectors.)

1.1.2 Definition. If the vector space $\mathbb{F}$ over $\mathbb{E}$ is finite-dimensional its dimension is denoted by $[\mathbb{F} : \mathbb{E}]$ and called the *degree of* $\mathbb{F}$ *over* $\mathbb{E}$. ■

For example, it is readily seen that $[\mathbb{C} : \mathbb{R}] = 2$ and $[\mathbb{R} : \mathbb{R}] = 1$.

Exercises 1.1

1. (a) Which of the following are meaningful?
   (i) the vector space $\mathbb{C}$ over $\mathbb{R}$;
   (ii) the vector space $\mathbb{R}$ over $\mathbb{R}$;
   (iii) the vector space $\mathbb{R}$ over $\mathbb{C}$;
   (iv) the vector space $\mathbb{C}$ over $\mathbb{Q}$;
   (v) the vector space $\mathbb{C}$ over $\mathbb{C}$;
   (vi) the vector space $\mathbb{Q}$ over $\mathbb{Q}$;
   (vii) the vector space $\mathbb{Q}$ over $\mathbb{R}$;
   (viii) the vector space $\mathbb{Q}$ over $\mathbb{C}$.

   (b) Which of the above vector spaces has dimension 2? Write down a basis in each such case.

2. How may subfields of $\mathbb{Q}$ are there? (Justify your answer.)

3. (a) Is $\{(1 + i\sqrt{2}), (\sqrt{2} + 2i)\}$ a linearly independent subset of the vector space $\mathbb{C}$ over $\mathbb{R}$?
   (b) Is $\{(1 + i\sqrt{2}), (\sqrt{2} + 2i)\}$ a linearly independent subset of the vector space $\mathbb{C}$ over $\mathbb{Q}$?

4. (a) Let $M_2(\mathbb{R})$ be the set of all 2×2 matrices with entries from $\mathbb{R}$. What are the natural vector space addition and scalar multiplication which make $M_2(\mathbb{R})$ a vector space over $\mathbb{R}$?
   (b) Find a basis for $M_2(\mathbb{R})$.
   (c) Let $\mathbb{M}_2(\mathbb{C})$ be the set of all 2×2 matrices with entries from $\mathbb{C}$. Show that it is a vector space over $\mathbb{R}$ and find a basis for it. Is it also a vector space over $\mathbb{C}$?

5. Let $\mathbb{F} = \{a + b\sqrt{2};\, a, b \in \mathbb{Q}\}$.

(a) Prove that $\mathbb{F}$ is a vector space over $\mathbb{Q}$ and write down a basis for it.

(b)* Prove that $\mathbb{F}$ is a field.

(c) Find the value of $[\mathbb{F} : \mathbb{Q}]$.

6. A ring R is said to be an *integral domain* if its multiplication is commutative, if there is an element 1 such that $1.x = x.1 = x$ for all $x \in R$, and if there are no zero divisors.
   (a) Is $\mathbb{Z}_4$ an integral domain?

(b)* For what $n \in \mathbb{N}$ is $\mathbb{Z}_n$ an integral domain?

7. Let $\mathbb{E}$ be a subfield of a finite field $\mathbb{F}$ such that $[\mathbb{F} : \mathbb{E}] = n$, for some $n \in \mathbb{N}$. If $\mathbb{E}$ has m elements, for some $m \in \mathbb{N}$, how many elements does $\mathbb{F}$ have? (Justify your answer.)

8. (a) If A, B, and C are finite fields with A a subfield of B and B a subfield of C, prove that $[C : A] = [C : B].[B : A]$.
   [Hint. Use the result of Exercise 7.]
   (b) If furthermore $[C : A]$ is a prime number, deduce that $B = C$ or $B = A$.

1.2 Polynomials

The simplest and most naive way of describing a polynomial is to say that it is an "expression" of the form

$$a_0 + a_1x + a_2x^2 + \ldots + a_nx^n. \tag{1}$$

The coefficients $a_0, a_1, \ldots, a_n$ are assumed to belong to some ring R. The above expression is then called a *polynomial over the ring R*.

As yet we have said nothing about the symbol x appearing in the above expression. There are, in fact, two distinct ways of interpreting this symbol. One of these ways leads to the concept of a *polynomial form*, the other to that of a *polynomial function*.

Polynomial Forms

To arrive at the concept of a polynomial form, we say as little as possible about the symbol "x". We regard x and its various powers as simply performing the rôle of "markers" to indicate the position of the various coefficients in the expression.

Thus, for example, we know that the two polynomials

$$1 + 2x + 3x^2 \quad \text{and} \quad 2x + 3x^2 + 1$$

are equal because in each expression the coefficients of the same powers of x are equal.

Again, when we add two polynomials, as in the following calculation,

$$(1 + 2x + 3x^2) + (2 + x^2) = 3 + 2x + 4x^2,$$

we may regard the powers of x as simply telling us which coefficients to add together. They play a similar, although more complicated, rôle when we multiply two polynomials.

Although x is regarded as obeying the same algebraic rules as the elements of the ring R, we do not think of it as assuming values from R. For this reason it is called an *indeterminate*.

For emphasis, we shall use upper case letters for indeterminates in this book and write "X" instead of "x" so our polynomial (1) becomes

$$a_0 + a_1X + a_2X^2 + \ldots + a_nX^n,$$

and we call it a *polynomial form* (over the ring R, in the indeterminate X).

What really matters in a polynomial form are the coefficients. Accordingly we say that *two polynomial forms are equal if and only if they have the same coefficients.*

1.2.1 Definitions. If in the above polynomial form the last coefficient $a_n \neq 0$, then we say that the polynomial has *degree* n.

If all the coefficients are zero, we say that the polynomial is the *zero polynomial* and leave its degree undefined. ■

When we do not wish to state the coefficients explicitly we shall use symbols like $f(X), g(X), h(X)$ to denote polynomials in X. The degree of a nonzero polynomial $f(X)$ is denoted by

$$\deg f(X).$$

If in the above polynomial form $a_n \neq 0$, we call a_n the *leading coefficient*.

The collection of all polynomials over the ring R in the indeterminate X will be denoted by

$$R[X]$$

(note the square brackets, as distinct from round ones). Since polynomial forms can be added, subtracted and multiplied (and these operations obey the usual algebraic laws) this set of polynomials is itself a ring.

In particular, if the ring R happens to be a field $\mathbb{F}$, we obtain the polynomial ring $\mathbb{F}[X]$. This ring will not be a field, however, since $X \in \mathbb{F}[X]$ but X has no reciprocal in $\mathbb{F}[X]$ because

$$X.f(X) \neq 1, \quad \text{for all} \quad f(X) \in \mathbb{F}[X].$$

Polynomial Functions

Back in the original expression (1), an alternative way to regard the symbol "x" is as a variable standing for a typical element of the ring R. Thus the expression (1) may be used to assign to each element $x \in R$ another element in R. In this way we get a function $f : R \to R$ with values assigned by the formula

$$f(x) = a_0 + a_1x + a_2x^2 + \ldots + a_nx^n. \tag{2}$$

Such a function f is called a *polynomial function* on the ring R.

Thus if we regard polynomials as functions, emphasis shifts from the coefficients to the values of the function. In particular, *equality of two polynomial functions* $f : R \to R$ and $g : R \to R$ means that

$$f(x) = g(x), \qquad \text{for all} \quad x \in R,$$

which is just the standard definition of equality for functions.

Clearly each polynomial form $f(X)$ determines a unique polynomial function $f : R \to R$ (because we can read off the coefficients from $f(X)$ and use them to generate the values of f via the formula (2)). Is the converse true? *Does each polynomial function $f : R \to R$ determine a unique polynomial form?* The answer is: *not necessarily!* The following example shows why.

1.2.2 Example. Let $f(X)$ and $g(X)$ be the polynomial forms over the ring $\mathbb{Z}_4$ given by

$$f(X) = 2X \quad \text{and} \quad g(X) = 2X^2.$$

These two polynomial forms are different, yet they determine the same polynomial function.

Proof. These polynomial forms determine two functions $f : \mathbb{Z}_4 \to \mathbb{Z}_4$ and $g : \mathbb{Z}_4 \to \mathbb{Z}_4$, with values given by

$$f(x) = 2x \quad \text{and} \quad g(x) = 2x^2.$$

Hence calculation in $\mathbb{Z}_4$ shows that

$$\begin{aligned} f(0) &= 0 = g(0) \\ f(1) &= 2 = g(1) \\ f(2) &= 0 = g(2) \\ f(3) &= 2 = g(3). \end{aligned}$$

So $f(x) = g(x)$, for all $x \in \mathbb{Z}_4$. Thus the functions f and g are equal. ■

This is something of an embarrassment! We have produced an example of a ring R and two polynomial forms $f(X), g(X) \in R[X]$ such that

$$f(X) \neq g(X) \quad \text{and yet} \quad f = g.$$

Fortunately, however, the only rings which are relevant to our geometrical construction problems are those which are subfields of the complex number field, $\mathbb{C}$. For such rings it can be shown (see Exercises 1.2 #6) that the above phenomenon cannot occur and we get a one-to-one correspondence

$$f(X) \leftrightarrow f$$

between polynomial forms $f(X) \in R[X]$ and polynomial functions $f : R \to R$.

The result of all this is that while there is a fine conceptual distinction between polynomial forms and polynomial functions, we can ignore the distinction in the remainder of this book without running into practical difficulties. *Thus*

$$f(X) \quad \textit{and} \quad f$$

will be more or less the same while $f(x)$ *will be quite different, being the value of* f *at* x.

Exercises 1.2

1. Give an example of an element of the polynomial ring $\mathbb{R}[X]$ which is not a member of $\mathbb{Q}[X]$.

2. In each of the following cases, give a pair of nonzero polynomials $f(X), g(X) \in \mathbb{Q}[X]$ which satisfies the condition:
   (i) $\deg(f(X) + g(X)) < \deg f(X) + \deg g(X)$;
   (ii) $\deg(f(X) + g(X)) = \deg f(X) + \deg g(X)$;
   (iii) $\deg(f(X) + g(X))$ is undefined.

3. (a) Let $\mathbb{F}$ be a field. Verify that if $f(X)$ and $g(X)$ are nonzero polynomials in $\mathbb{F}[X]$ then $f(X)g(X) \neq 0$ and

$$\deg f(X)g(X) = \deg f(X) + \deg g(X).$$

   (b) Deduce that the ring $\mathbb{F}[X]$ is an integral domain.
   (See Exercises 1.1 #6 for the definition of integral domain.)
   (c) Is $\mathbb{Z}_6[X]$ an integral domain? (Justify your answer.)

4. Let $f(X)$ denote the polynomial form over the ring $\mathbb{Z}_6$ given by $f(X) = X^3$. Find a polynomial form $g(X)$, with $f(X) \neq g(X)$, such that $f(X)$ and $g(X)$ determine the same polynomial function.

5.* Let R be any finite ring. Prove that there exist unequal polynomial forms $f(X)$ and $g(X)$ over R such that $f(X)$ and $g(X)$ determine the same polynomial function.

6. Let $\mathbb{F}$ be any subfield of the complex number field $\mathbb{C}$ and let $f(X)$ and $g(X)$ be polynomial forms over the field $\mathbb{F}$. If the polynomial functions f and g are equal (that is, $f(x) = g(x)$, for all $x \in \mathbb{F}$), prove that the polynomial forms are equal (that is, $f(X) = g(X)$).
   [Hint. You may assume the well-known result that for any polynomial function h of degree n, there are at most n complex numbers x such that $h(x) = 0$.]

1.3 The Division Algorithm

When you divide one polynomial by another you get a quotient and a remainder. The procedure for doing this is called the *division algorithm* for polynomials. It leads to a theorem which is of fundamental importance in the study of polynomials. We begin with an example.

1.3.1 Example. We shall divide the polynomial $X^3 + 2X^2 + 3X + 4$ by the polynomial $X + 1$ to produce a quotient and remainder in $\mathbb{Q}[X]$.

$$\begin{array}{rl}
 & X^2 + X + 2 \qquad \leftarrow \textit{quotient} \\
\textit{divisor} \rightarrow X + 1 & \overline{\big)\, X^3 + 2X^2 + 3X + 4} \quad \leftarrow \textit{dividend} \\
 & \underline{X^3 + X^2} \\
 & \quad X^2 + 3X + 4 \\
 & \quad \underline{X^2 + X} \\
 & \qquad 2X + 4 \\
 & \qquad \underline{2X + 2} \\
 & \qquad\quad 2 \leftarrow \textit{remainder}
\end{array}$$

It is easy to check that the algorithm has done its job and given us what we want, namely,

$$\textit{dividend} = \textit{divisor} \times \textit{quotient} + \textit{remainder}.$$

To understand why the division algorithm works it is helpful to write the results from the various stages of the process as equalities:

$$\left.\begin{aligned}
X^3 + 2X^2 + 3X + 4 &= (X + 1)X^2 + (X^2 + 3X + 4) \\
X^2 + 3X + 4 &= (X + 1)X + (2X + 4) \\
2X + 4 &= (X + 1)2 + 2
\end{aligned}\right\} \qquad (1)$$

To help understand the significance of the equations (1), let us introduce some names for the various terms occurring in them and hence write each of the Eqs. (1) as

$$\textit{current dividend} = (X + 1)(\textit{monomial in } X) + \textit{current remainder}.$$

(A *monomial* in X means a polynomial of the type cX^i.) In the first of the equations (1), the current dividend is the actual dividend. At each step, the aim is to choose the monomial so as to balance out the highest power of X in the current dividend. This will ensure that

$$\deg \textit{current remainder} < \deg \textit{current dividend}.$$

At the next step, the old current remainder becomes the new current dividend. The process stops when eventually (at the third stage in our example) the degree of the current remainder falls below the degree of the divisor ($X + 1$ in this case).

The final step is to put the three equations (1) together to get

$$\underbrace{X^3 + 2X^2 + 3X + 4}_{\textit{dividend}} = \underbrace{(X + 1)}_{\textit{divisor}}\underbrace{(X^2 + X + 2)}_{\textit{quotient}} + \underbrace{2}_{\textit{remainder}}$$

thereby verifying that the algorithm has achieved its goal. ■

The division algorithm can be applied more generally to a dividend $f(X)$ and divisor $g(X)$ to give a quotient $q(X)$ and remainder $r(X)$ as in the following theorem.

1.3.2 Theorem. [Division Theorem] *Let $\mathbb{F}$ be a field. If $f(X)$ and $g(X)$ are in $\mathbb{F}[X]$ and $g(X) \neq 0$, then there are polynomials $q(X)$ and $r(X)$ in $\mathbb{F}[X]$ such that*

$$f(X) = g(X)q(X) + r(X)$$

and either $r(X) = 0$ or $\deg r(X) < \deg g(X)$.

Proof. The division algorithm is used to construct the polynomials $q(X)$ and $r(X)$. The proof that the algorithm succeeds uses the same ideas as those explained in the above example. ■

As an immediate corollary of the Division Theorem 1.3.2, we obtain:

1.3.3 Theorem. [Remainder Theorem] *Let $f(X)$ and $g(X)$ be as in the Division Theorem 1.3.2. If $g(X) = X - c$, for $c \in \mathbb{F}$, then the $r(X)$ in that theorem is $f(c)$.*

When the $r(X)$ in the Division Theorem 1.3.2 equals $0 \in \mathbb{F}$, we say that the polynomial $f(X)$ is *divisible* by the polynomial $g(X)$.

Exercises 1.3

1. Find the quotient and remainder when $X^3 + 2X + 1$ is divided by $2X + 1$ in $\mathbb{Q}[X]$.
2. (a) Repeat Exercise 1 above with $\mathbb{Z}_7[X]$ replacing $\mathbb{Q}[X]$.

   (b) Write your answer to (a) without using any "–" signs.

3. Use the division algorithm to find the quotient and remainder when $X^3 + iX^2 + 3X + i$ is divided by $X^2 - i$ in $\mathbb{C}[X]$.
4. Use the division algorithm to find the quotient and remainder when $X^2 + \sqrt{2}X - 3$ is divided by $X - \sqrt{3}$ in $\mathbb{R}[X]$.
5.* Write out the proof of Theorem 1.3.2 in detail.
[Hint. Your proof may use mathematical induction.]
6.* Find an example which shows that Theorem 1.3.2 would be false if we assumed that $\mathbb{F}$ was only a ring rather than a field.
7.* In Theorem 1.3.2 if $\mathbb{F}$, $f(X)$ and $g(X)$ are given, show that $q(X)$ and $r(X)$ are uniquely determined.

1.4 The Rational Roots Test

This simple test will enable you to prove very easily that certain numbers, such as

$$\sqrt{3}, \quad \sqrt[5]{2}, \quad \text{and} \quad \sqrt{2}+\sqrt{3},$$

are irrational. The test illustrates how polynomials can be applied to the study of numbers.

A *zero* of a polynomial $f(X)$ is a number β such that $f(\beta) = 0$. The test works by narrowing down to a short list the possible zeros in $\mathbb{Q}$ of a polynomial in $\mathbb{Q}[X]$.

A preliminary observation is that every polynomial in $\mathbb{Q}[X]$ can be written as a rational multiple of a polynomial in $\mathbb{Z}[X]$. This is achieved by multiplying the given polynomial in $\mathbb{Q}[X]$ by a suitable integer. For example

$$1 + \frac{1}{3}X + \frac{2}{7}X^2 + \frac{1}{2}X^3 = \frac{1}{42}(42 + 14X + 12X^2 + 21X^3).$$

In looking for zeros in $\mathbb{Q}$ of polynomials in $\mathbb{Q}[X]$ we may as well, therefore, look for zeros in $\mathbb{Q}$ of polynomials in $\mathbb{Z}[X]$.

The Rational Roots Test uses some terminology which we now record.

By saying that an integer p *is a factor of* an integer q, we mean that there is an integer m such that $q = mp$.

By saying that a rational number β is expressed in *lowest terms* by $\beta = \frac{r}{s}$ we mean that r and s are in $\mathbb{Z}$ with $s \neq 0$ and r and s have no common factors except 1 and -1.

1.4.1 Theorem. [Rational Roots Test] *Let $f(X) \in \mathbb{Z}[X]$ be a polynomial of degree n so that*

$$f(X) = a_0 + a_1X + \ldots + a_nX^n$$

for some $a_0, a_1, \ldots, a_n \in \mathbb{Z}$, with $a_n \neq 0$. If β is a rational number, written in its lowest terms as $\beta = \frac{r}{s}$, and β is a zero of $f(X)$ then

(i) r is a factor of a_0, and
(ii) s is a factor of a_n.

Proof. Let $f(\beta) = 0$ so that

$$a_0 + a_1\left(\frac{r}{s}\right) + a_2\left(\frac{r}{s}\right)^2 + \ldots + a_n\left(\frac{r}{s}\right)^n = 0$$

and hence

$$a_0 s^n + a_1 r s^{n-1} + a_2 r^2 s^{n-2} + \ldots + a_n r^n = 0.$$

This gives

$$a_0 s^n = -r\left(a_1 s^{n-1} + a_2 r s^{n-2} + \ldots + a_n r^{n-1}\right)$$

so that

$$r \quad \text{is a factor of} \quad a_0 s^n.$$

But since r and s have no common factors except 1 and -1, there can be no common prime numbers in the prime factorizations of r and s. This implies that

$$r \quad \text{is a factor of} \quad a_0.$$

(This follows from the Fundamental Theorem of Arithmetic; see Exercises 1.4 #9.)

Similarly on writing

$$a_n r^n = -s\left(a_0 s^{n-1} + a_1 r s^{n-2} + \ldots + a_{n-1} r^{n-1}\right)$$

we see that

$$s \quad \text{is a factor of} \quad a_n r^n$$

and therefore

$$s \quad \text{is a factor of} \quad a_n.$$

■

The following example is a typical application of the Rational Roots Test.

1.4.2 Example. The real number $\sqrt[5]{2}$ is not rational.

Proof. Note firstly that $\sqrt[5]{2}$ is a zero of the polynomial $2 - X^5$ in $\mathbb{Q}[X]$. This polynomial also lies in $\mathbb{Z}[X]$, and hence we can use the Rational Roots Test to see if it has a zero in $\mathbb{Q}$.

Suppose therefore that $\frac{r}{s}$ is a zero of $2 - X^5$, where $r, s \in \mathbb{Z}$ with $s \neq 0$ and $\frac{r}{s}$ is expressed in lowest terms. By the Rational Roots Test,

$$r \text{ is a factor of } 2 \quad \text{and} \quad s \text{ is a factor of } -1.$$

This means that the only possible values of r are $1, -1, 2$, and -2, while the only possible values of s are 1 and -1. Hence $\frac{r}{s}$ must be $1, -1, 2$, or -2. This means that ± 1 and ± 2 are the only possible zeros in $\mathbb{Q}$. Substitution shows, however, that none of ± 1 and ± 2 is a zero of $2 - X^5$, and so $2 - X^5$ has no zeros in $\mathbb{Q}$. Hence $\sqrt[5]{2}$, being a zero, is not in $\mathbb{Q}$. ■

Exercises 1.4

1. Let $p(X)$ be the element of $\mathbb{Q}[X]$ given by

$$p(X) = 2X^3 + 3X^2 + 2X + 3.$$

   (a) Use the Rational Roots Test to find all possible rational zeros of $p(X)$.
   (b) Is -1 a zero of $p(X)$?
   (c) Is $-\frac{3}{2}$ a zero of $p(X)$?

2. Use the Rational Roots Test to prove that $\sqrt{5}$ is irrational.

3. Find a polynomial in $\mathbb{Q}[X]$ which has $\sqrt{2} + \sqrt{3}$ as a zero. Hence show that $\sqrt{2} + \sqrt{3}$ is irrational.
   [Hint. To obtain the polynomial, firstly let $\alpha = \sqrt{2} + \sqrt{3}$ and then square both sides.]

4. (a) Use the Rational Roots Test to prove that for each $m \in \mathbb{N}$, $\sqrt{m}$ is rational if and only if m is a perfect square.
   (b) Let m and n be any positive integers. Prove that $\sqrt[n]{m}$ is a rational number if and only if it is an integer.

5. If n is any positive integer, prove that $\sqrt{n} + \sqrt{n+1}$ and $\sqrt{n} - \sqrt{n+1}$ are irrational.

6. Prove that $\sqrt{n+1} - \sqrt{n-1}$ is irrational, for every positive integer n.

7.* Prove the following numbers are irrational:

(a) $\sqrt{3} + \sqrt{5} + \sqrt{7}$, and
(b) $\sqrt{2} + \sqrt{3} + \sqrt{5}$.
   [Hint. Both (a) and (b) involve some arithmetic. Do not be afraid to use a calculator or a computer program in an appropriate fashion.]

8. Find a polynomial in $\mathbb{Q}[X]$ with α as a zero, where

$$\alpha = \sqrt[3]{4} - \sqrt[3]{2}$$

   and then deduce that $\alpha \notin \mathbb{Q}$ by applying the Rational Roots Test.
   [Hint. Calculate α^3 and keep an eye on α at the same time.]

9. *The Fundamental Theorem of Arithmetic* says that every natural number $n > 1$ can be written as a product of prime numbers and, except for the order of the factors, the expression of n in this form is unique. Thus n has a unique expression

$$n = p_1^{a_1} p_2^{a_2} \dots p_k^{a_k}$$

where $p_1, p_2, \dots, p_k$ are distinct prime numbers and $a_1, a_2, \dots, a_k$ are positive integers.

   (a) Use the Fundamental Theorem of Arithmetic to prove that if r and s are natural numbers such that a prime p is a factor of the product rs, then p must be a factor of r or of s (or of both).
   [Hint. Express r and s as products of primes and then use the uniqueness part of the Fundamental Theorem.]
   (b) Let r, s, and b be positive integers such that r and s have no common factors except 1 and -1. Use the Fundamental Theorem of Arithmetic to show that if r is a factor of bs, then r must be a factor of b.
   (c) Let r, s, b, and m be positive integers such that r and s have no common factors except 1 and -1. Show that if r is a factor of bs^m, then r must be a factor of b.

10.* Use the Rational Roots Test to prove that the number $\sin\frac{\pi}{9}$ (that is, $\sin 20°$) is irrational.
[Hint. Firstly, use the formulae

$$\cos^2\theta + \sin^2\theta = 1,$$
$$\sin(\theta+\phi) = \sin\theta\cos\phi + \cos\theta\sin\phi,$$
$$\cos(\theta+\phi) = \cos\theta\cos\phi - \sin\theta\sin\phi.$$

to write $\sin 3\theta$ in terms of $\sin\theta$ (without any cos terms). Next, put $\theta = \frac{\pi}{9}$, $x = \sin\theta$, and use $\sin\frac{\pi}{3} = \frac{\sqrt{3}}{2}$.]

Appendix to Chapter 1

We record here formal definitions of ring, field, vector space, and of related terms. If you are not familiar with these concepts, we recommend that you read (Fraleigh, 1982, §23, §24 and §36), (Tropper, 1969, Chapters 1 & 2), or (Clark, 1971, the early parts of Chapter 3). The books Gallian (2017) and Gouvêa (2012) have a good coverage of groups, rings, and fields and Gallian (2017) also has a historical discussion of geometric constructions. Adamson (2007) is a nice introduction to field theory and discusses ruler and compass constructions.

1.A.1 Definitions. A *group* $(G, +)$ is a set G together with a binary operation $+$ such that

(a) if $g_1, g_2 \in G$, then $g_1 + g_2 \in G$;
(b) if $g_1, g_2, g_3 \in G$, then $g_1 + (g_2 + g_3) = (g_1 + g_2) + g_3$;
(c) there is an element $0 \in G$ such that $g + 0 = 0 + g = g$ for all $g \in G$;
(d) for every element $g \in G$, there exists an element g' such that $g+g'=g' + g = 0$.

The element 0 is called the *identity* of the group G. The element g' in (d) is called the *inverse* of g.

The group G is said to be *abelian* or *commutative* if $g_1 + g_2 = g_2 + g_1$ for all $g_1, g_2 \in G$. ■

We shall be more interested in sets with two binary operations.

1.A.2 Definitions. A *ring* is a set R with two binary operations $+$ and $\cdot$ such that

(a) R together with the operation $+$ is an abelian group;
(b) if $r_1, r_2 \in R$, then $r_1.r_2 \in R$;
(c) if $r_1, r_2, r_3 \in R$, then $r_1 \cdot (r_2 \cdot r_3) = (r_1 \cdot r_2) \cdot r_3$;
(d) if $r_1, r_2, r_3 \in R$, then

$$r_1 \cdot (r_2 + r_3) = r_1 \cdot r_2 + r_1 \cdot r_3 \quad \text{and} \quad (r_2 + r_3) \cdot r_1 = r_2 \cdot r_1 + r_3 \cdot r_1.$$

The additive identity of R is denoted by 0.

The ring R is said to be a *commutative ring* if $r_1 \cdot r_2 = r_2 \cdot r_1$ for all $r_1, r_2 \in R$.

A *ring with unity* is a ring R with an element 1 (called the *unity* of R) such that $1 \cdot r = r \cdot 1 = r$ for all $r \in R$.

An element r of a ring $(R, +, \cdot)$ is said to be a *zero divisor* if $r \neq 0$ and there exists an element $s \neq 0$ in R such that $r \cdot s = 0$ or $s \cdot r = 0$.

A ring $(R, +, \cdot)$ is said to be an *integral domain* if it is a commutative ring with unity and it contains no zero divisors.

A subset S of a ring $(R, +, \cdot)$ is said to be a *subring* of R if $(S, +, \cdot)$ is itself a ring. ■

1.A.3 Definitions. A ring $(\mathbb{F}, +, \cdot)$ is said to be a *field* if $\mathbb{F}\backslash\{0\}$ together with the operation $\cdot$ is an abelian group.

A subset S of a field (F, $+, \cdot$) is said to be a *subfield* of $\mathbb{F}$ if $(S, +, \cdot)$ is itself a field. ■

The additive identity of a field $\mathbb{F}$ is usually denoted by 0, and the multiplicative identity of $\mathbb{F}$ is denoted by 1.

Note that while every field is an integral domain, there are integral domains which are not fields: for example, the ring $\mathbb{Z}$ of integers is an integral domain which is not a field.

1.A.4 Definitions. A set V together with two operations $+$ and $\cdot$ together with a field $\mathbb{F}$ is said to be a *vector space over the field* $\mathbb{F}$ if

(a) V together with the operation $+$ is an abelian group;
(b) if $\lambda \in \mathbb{F}$ and $v \in V$, then $\lambda \cdot v \in V$;
(c) if $\lambda_1, \lambda_2 \in \mathbb{F}$ and $v \in V$, then $(\lambda_1 + \lambda_2) \cdot v = \lambda_1 \cdot v + \lambda_2 \cdot v$;
(d) if $v_1, v_2 \in V$ and $\lambda \in \mathbb{F}$, then $\lambda \cdot (v_1 + v_2) = \lambda \cdot v_1 + \lambda \cdot v_2$;
(e) if 1 is the multiplicative identity of $\mathbb{F}$, then $1 \cdot v = v$ for all $v \in V$;
(f) if $\lambda_1, \lambda_2 \in \mathbb{F}$ and $v \in V$, then $(\lambda_1 \lambda_2) \cdot v = \lambda_1 \cdot (\lambda_2 \cdot v)$.

We refer to the elements of $\mathbb{F}$ as the *scalars* and the multiplication of an element of $\mathbb{F}$ by an element of V (for example, $\lambda.v$ in (b) above) as *scalar multiplication*. The elements of V are called *vectors*. The identity of the group $(V, +)$ is called the *zero vector* and is denoted by 0. The operation $+$ is often called *vector addition*.

A subset S of the vector space V over the field $\mathbb{F}$ is said to be a *subspace* of V if S together with the operations $+$ and $\cdot$ of V is a vector space over the field $\mathbb{F}$. ■

1.A.5 Definitions. If V is a vector space over a field $\mathbb{F}$ and S is a subset of V, then the set

$$\text{span}(S) = \{\lambda_1 \cdot v_1 + \ldots + \lambda_n \cdot v_n : n \geq 1 \text{ and each } \lambda_i \in \mathbb{F},\ v_i \in S\}$$

is called the *span* of S over $\mathbb{F}$ or the *linear span* of S over $\mathbb{F}$.

If span $(S) = V$ then S is said to *span* V over $\mathbb{F}$. ■

1.A.6 Definitions. If V is a vector space over a field $\mathbb{F}$ and S is a subset of V, then S is said to be a *linearly independent set over* $\mathbb{F}$ if the zero vector, 0, can be written as a linear combination of vectors in S in only one way, namely with all the coefficients equal to 0; that is, for $v_1, v_2, \ldots, v_n \in S$,

$$0 = \lambda_1 \cdot v_1 + \lambda_2 \cdot v_2 + \ldots + \lambda_n \cdot v_n$$

implies that $\lambda_1 = \lambda_2 = \ldots = \lambda_n = 0$.

S is said to be *linearly dependent over* $\mathbb{F}$ if it is not linearly independent over $\mathbb{F}$. ■

1.A.7 Definitions. If V is a vector space over a field $\mathbb{F}$ and S is a nonempty subset of V which is linearly independent over $\mathbb{F}$ and which spans V over $\mathbb{F}$, then S is said to be a *basis of* V *over* $\mathbb{F}$. If S is finite, then the number of elements of S is called the *dimension of* V *over* $\mathbb{F}$. ■

Additional Reading for Chapter 1

General algebra books which include the background information we assume about rings, fields, and vector spaces (and which also deal with geometrical constructions) include Clark (1971), Fraleigh (1982), Gallian (2017), Gouvêa (2012), Sethuraman (1997), and Shapiro (1975). Tropper (1969) is a good introduction to vector spaces and Adamson (2007) is an introduction to field theory with a discussion of ruler and compass constructions. Deeper results about irrational numbers can be found in Hardy and Wright (1960) (which is the classic reference on number theory) and in Niven (1961) and Niven (1967). Hardy and Wright (1960) contains a proof of the Fundamental Theorem of Arithmetic and is also a good reference on prime numbers. Lang (2002) is a graduate text covering, in particular, groups, rings, fields and polynomials.

Chapter 2
Algebraic Numbers and Their Polynomials

Straightedge and compass constructions can be used to produce line segments of various lengths relative to some preassigned unit length. Although each of these lengths is a real number, it turns out that not every real number can be obtained in this way. The lengths which can be constructed are rather special.

As the first step towards classifying the lengths which can be constructed, this chapter introduces the concept of an algebraic number (or more specifically of a number which is algebraic over a field). Each such number will satisfy many polynomial equations and our immediate goal is to choose the simplest one.

Over 1,500 years ago Chinese, Greek, and Indian mathematicians were familiar with the number π as the ratio of the circumference of a circle to its diameter and were able to approximate π correctly to a few decimal places. (See Berggren et al. (1997).) The Swiss polymath Johann Heinrich Lambert (1728–1777) proved in 1761 that π is an irrational number; that is, it is not the ratio of two integers. This was in fact conjectured by the Greek philosopher and polymath Aristotle some 2,000 years earlier. The symbol π for this number appeared in the book Jones (2018, first published 1706), by the Welsh mathematician William Jones (1675–1749), who chose the first letter of the Greek word $\pi\varepsilon\rho\iota\varphi\varepsilon\rho\epsilon\iota\alpha$ meaning periphery or circumference. The number e, which is the base of the natural logarithms, was introduced in 1683 by the Swiss mathematician Jacob Bernoulli (1655–1705) in his study of compound interest. The symbol e for this number first appeared in a paper by the Swiss polymath Leonhard Euler (1707–1783) who proved in 1737 that e is irrational. Euler introduced the notion of a transcendental number and Lambert conjectured that both e and π are transcendental. In 1873 the French mathematician Charles Hermite (1822–1901) proved that e is a transcendental number. In 1882 the German mathematician Ferdinand von Lindemann (1852–1939) proved the beautiful result that π also is a transcendental number.

S. A. Morris et al. *Abstract Algebra and Famous Impossibilities*, Undergraduate Texts in Mathematics, https://doi.org/10.1007/978-3-031-05698-7_2

2.1 Algebraic Numbers

Numbers which lie in $\mathbb{R}$ but not in $\mathbb{Q}$ are said to be *irrational*. Well-known examples are $\sqrt{2}$, e and π. The irrationality of $\sqrt{2}$ was known to the ancient Greeks whereas that of e and π was proved hundreds of years later. Although these three numbers are all irrational, there is a fundamental distinction between $\sqrt{2}$ and the other two numbers. While $\sqrt{2}$ satisfies a polynomial equation

$$x^2 - 2 = 0$$

with coefficients in $\mathbb{Q}$, no such equation is satisfied by e or π (which we shall prove in Chapter 10.) For this reason $\sqrt{2}$ is said to be an *algebraic number* whereas e andπ are said to be *transcendental numbers*.

The precise definition is as follows.

2.1.1 Definition. A number $\alpha \in \mathbb{C}$ is said to be *algebraic over a field* $\mathbb{F} \subseteq \mathbb{C}$ if there is a nonzero polynomial $f(X) \in \mathbb{F}[X]$, such that α is a zero of $f(X)$; that is, there is a polynomial

$$f(X) = a_0 + a_1 X + \ldots + a_n X^n$$

whose coefficients $a_0, a_1, \ldots, a_n$ all belong to $\mathbb{F}$, at least one of these coefficients is nonzero, and $f(\alpha) = 0$. ■

Observe that for each field $\mathbb{F}$, every number α in $\mathbb{F}$ is algebraic over $\mathbb{F}$ because α is a zero of the polynomial $X - \alpha \in \mathbb{F}[X]$. This implies that e and π are algebraic over $\mathbb{R}$ even though they are not algebraic over $\mathbb{Q}$, as noted above.

2.1.2 Examples.

(i) The number $\sqrt{2}$ is algebraic over $\mathbb{Q}$ because it is a zero of the polynomial $X^2 - 2$, which is nonzero and has coefficients in $\mathbb{Q}$.
(ii) The number $\sqrt[4]{2}\sqrt[6]{3}$ is algebraic over $\mathbb{Q}$ because it is a zero of the polynomial $X^{12} - 72$, which is nonzero and has coefficients in $\mathbb{Q}$. ■

In order to show that a number is algebraic, we look for a suitable polynomial having that number as a zero.

2.1.3 Example. The number $1 + \sqrt{2}$ is algebraic over $\mathbb{Q}$.

Proof. Let $\alpha = 1 + \sqrt{2}$.

We want to find a polynomial, with α as a zero, which has coefficients in $\mathbb{Q}$. This suggest squaring to get rid of the square roots:

$$\alpha^2 = 1 + 2\sqrt{2} + 2,$$

which in no good as $\sqrt{2}$ still appears on the right hand side. So we go back and isolate $\sqrt{2}$ on one side before squaring!

Isolating $\sqrt{2}$ gives $\quad \alpha - 1 = \sqrt{2}.$
Squaring both sides gives $\quad (\alpha - 1)^2 = 2,$
and so $\quad \alpha^2 - 2\alpha - 1 = 0.$

Thus α is a zero of the polynomial $X^2 - 2X - 1$, which is nonzero and has coefficients in $\mathbb{Q}$. Hence α is algebraic over $\mathbb{Q}$. ■

2.1.5 Other Versions.

It is useful to be able to recognize the definition of "algebraic over a field $\mathbb{F}$" when it appears in different guises: thus $\alpha \in \mathbb{C}$ *is algebraic over* $\mathbb{F}$ *if and only if*
(*i*) *there exists* $f(X) \in \mathbb{F}[X]$ *such that* $f(X) \neq 0$ *and* $f(\alpha) = 0$, *OR*
(*ii*) *there is a positive integer* n *and numbers* $a_0, a_1, \ldots, a_{n-1}, a_n$ *in* $\mathbb{F}$, *not all zero, such that*

$$a_0 + a_1\alpha + a_2\alpha^2 + \cdots + a_{n-1}\alpha^{n-1} + a_n\alpha^n = 0.$$

The last statement may ring a few bells! Where have you seen it before? Well, to put things in context recall that, since $\mathbb{F}$ is a subfield of $\mathbb{C}$, *we can regard* $\mathbb{C}$ *as a vector space over* $\mathbb{F}$. (See the paragraph preceding Definition 1.1.2.) Now the numbers

$$1, \alpha, \alpha^2, \ldots, \alpha^{n-1}, \alpha^n$$

are all elements of $\mathbb{C}$, and hence can be regarded as vectors in the vector space $\mathbb{C}$ over $\mathbb{F}$. The coefficients $a_0, a_1, a_2, \ldots, a_{n-1}, a_n$, on the other hand, are all in $\mathbb{F}$ so we can regard them as scalars. Thus, we can write (ii) in the alternative way:

(*iii*) *there is a positive integer* n *such that the set of powers*

$$\{1, \alpha, \alpha^2, \ldots, \alpha^{n-1}, \alpha^n\}$$

of the number α *is* linearly dependent *over* $\mathbb{F}$. ■

You will often meet the terms “algebraic number” and “transcendental number” where no field is specified. In such cases the field is taken to be $\mathbb{Q}$. We formalize this below.

2.1.5 Definition. A complex number is said to be
(i) an *algebraic number* if it is algebraic over $\mathbb{Q}$;
(ii) a *transcendental number* if it is not algebraic over $\mathbb{Q}$. ■

Exercises 2.1

1. Verify that each of the following numbers is algebraic over the stated field:
   (a) $\sqrt{2}\sqrt[3]{5}$ over $\mathbb{Q}$;
   (b) $\sqrt{2}+\sqrt{3}$ over $\mathbb{Q}$;
   (c) $\sqrt{\pi}$ over $\mathbb{R}$;
   (d) i over $\mathbb{Q}$;
   (e) $\sqrt{2}+\sqrt{3}i$ over $\mathbb{R}$.
2. (a) Write down three different polynomials in $\mathbb{Q}[X]$, each of degree 2, which have $\sqrt{5}$ as a zero.
   (b) Write down three different polynomials in $\mathbb{Q}[X]$, each of degree greater than 2, which have $\sqrt{5}$ as a zero.
   (c) Write down three different polynomials in $\mathbb{Q}[X]$, each of degree 100, which have $\sqrt{5}$ as a zero.
3. Prove that if $\alpha \in \mathbb{C}$ is algebraic over $\mathbb{Q}$ then so is 2α.
4. Prove that if $\alpha \in \mathbb{C}$ is algebraic over a subfield F of $\mathbb{C}$ then so is $-\alpha$.
5. Prove that if α is a nonzero complex number which is algebraic over a subfield $\mathbb{F}$ of $\mathbb{C}$, then $1/\alpha$ is also algebraic over $\mathbb{F}$.
6. Show that every complex number is algebraic over $\mathbb{R}$.

7.* Let α be a positive real number, and let $\mathbb{F}$ be a subfield of $\mathbb{R}$. Prove that α is algebraic over $\mathbb{F}$ if and only if $\sqrt{\alpha}$ is algebraic over $\mathbb{F}$.

This next exercise shows you how to prove that π is an irrational number. The proof shall not be required in the material which follows.

This exercise is based on the 2003 New South Wales (a state in Australia) Mathematics Extension exam paper for the best mathematically inclined high school students. It assumes some calculus, specifically, integration by parts.

It is based on a proof in 1873 by the French mathematician Charles Hermite (1822–1901).

The proof outlined here is short and comparatively simple, but far from an obvious way to prove this result.

In Chapter 10 we shall prove the much stronger result that π is a transcendental number.

8.** Prove that π is an irrational number using the steps below.

We shall use proof by contradiction. So we begin by supposing that π is a rational number.

We shall define a sequence $I_0, I_1, I_2, \dots I_n \dots$ of real numbers. In fact, we shall prove that each I_n is a positive integer.

We shall also prove that the sequence $I_0, I_1, I_2, \dots I_n \dots$ tends to 0 as n tends to infinity.

But this is clearly impossible, as each positive integer must be at least 1 from 0 and so the sequence cannot tend to zero. Thus we shall have a contradiction, and hence π must be irrational.

(i) Suppose that π is not an irrational number. Then the positive number π must be given by $\pi = \dfrac{u}{v}$, for some positive integers u and v.

(ii) Define $I_0 = 2$ and $I_n = \dfrac{v^{2n}}{n!} \int_0^{\pi} x^n(\pi - x)^n \sin x \, dx$, for $n \in \mathbb{N}$.

(iii) Using integration by parts twice, verify that $I_1 = 4v^2$.

(iv) Verify that for $x \in [0, \pi]$, $x^n(\pi - x)^n \sin x \leq \pi^{2n}$.

(v) Deduce from (ii) and (iv) that $0 < I_n \leq \dfrac{v^{2n}}{n!}\pi^{2n+1}$.

(vi) Deduce from (v) that $I_n \to 0$ as $n \to \infty$.

(vii) Use integration by parts twice to show that

$$I_n = 2v^2(2n-1)I_{n-1} - u^2v^2I_{n-2}, \quad \text{for } n \geq 2.$$

(viii) As I_0, I_1, u, v are positive integers, deduce from (vii) that each I_n is an integer.

(ix) But (viii) and (v) say that each I_n is a positive integer, and (ii) says that the sequence of $I_0, I_1, \dots, I_n, \dots$ converges to zero, which is impossible. So we have a contradiction. Our supposition that π is a rational number is therefore false.

9. Use Exercise 8 above, which shows that π is an irrational number, to answer the following.
   (i) If r is any real number with $r \neq 0$, verify that at least one of the numbers $\pi + r$ and $\pi - r$ is an irrational number. Give an example where exactly one of these is irrational, and an example where both are irrational.
   (ii) If r is any real number with $r \neq 0$, verify that at least one of $\sqrt{\pi} \cdot r$ and $\frac{\sqrt{\pi}}{r}$ is an irrational number. Give an example where exactly one of these is irrational, and an example where both are irrational.
   (iii) If r is any real number with $r \neq 0$ and $r \neq \pi$, verify that at least two of the numbers $\pi^2 - r^2$, $\pi + r$, and $\pi - r$ are irrational numbers.
   [Hint. Use (i) and the fact that $\pi^2 - r^2 = (\pi - r)(\pi + r)$.]
   (iv) If r is any real number with $r \neq -\pi$, then at least one of the numbers $\frac{r}{\pi}$ and $r + \pi$ is an irrational number. Similarly show that if $r \neq \pi$, then at least one of $\frac{r}{\pi}$ and $r - \pi$ is an irrational number. Finally deduce that if $r \neq \pi$ and $r \neq -\pi$, then these imply that at least two of the numbers $\frac{r}{\pi}$, $r - \pi$, $r + \pi$ are irrational numbers.
   [Hint. Define $s = \frac{r}{\pi}$ and $t = r + \pi$. Eliminate r from the equations and solve for π in terms of s and t.]
   (v) Let r be any irrational number. Verify that at most one of the infinite set $\{m\pi + nr : m, n \in \mathbb{Z} \setminus \{0\},$ with m and n relatively prime$\}$ is a rational number.
   (vi) Verify that (i), (ii), (iii), (iv), and (v) remain true if π is replaced everywhere by any irrational number. (For example, we can replace π with $\sqrt{2}$.)
   (vii) Let $a, b, c \in \mathbb{R} \setminus \mathbb{Q}$, where a, b, and c are linearly independent over $\mathbb{Q}$. Prove that at most one of the 4 numbers $a + b + c$, $a + b - c$, $a - b + c$, $a - b - c$ is a rational number.

We remark that it is known that both Euler's constant, e, and π are irrational numbers, but it is not known if either $\pi - \mathrm{e}$ or $\pi + \mathrm{e}$ is an irrational number. But Exercise 9 (i) says that at least one of these two numbers is an irrational number.

2.2 Monic Polynomials

An algebraic number such as $\sqrt{2}$ will be a zero of many different polynomials. Ultimately we want to select from all these polynomials one which is, in some sense, the best. Here we make a start in this direction by restricting attention to **monic** polynomials, defined as follows:

2.2.1 Definition. A polynomial $f(X) = a_0 + a_1X + \cdots + a_nX^n$ in $\mathbb{F}[X]$ is said to be *monic* if its leading coefficient a_n is 1. ■

Thus, for example, $X^2 - 2$ is a monic polynomial whereas $3X^2 - 6$ is not. Note that both these polynomials have $\sqrt{2}$ as a zero and, for our purposes, the polynomial $X^2 - 2$ is somewhat nicer than $3X^2 - 6$. The following proposition shows that, in the definition of "algebraic over a field", there is no loss of generality in assuming that the polynomial is monic.

2.2.2 Proposition. *If a complex number α is a zero of a nonzero polynomial $f(X) \in \mathbb{F}[X]$, then α is a zero of a monic polynomial $g(X) \in \mathbb{F}[X]$ with* deg $g(X) =$ deg $f(X)$.

Proof. Assume that α is a zero of a polynomial $f(X) \neq 0$. Since $f(X) \neq 0$, some coefficient must be nonzero and (because there are only finitely many coefficients a_i) there must be a largest i such that $a_i \neq 0$. Let n be the largest such i.

Hence

$$f(X) = a_0 + a_1X + a_2X^2 + \ldots + a_nX^n$$

where $a_n \neq 0$. We choose

$$g(X) = \frac{1}{a_n} f(X).$$

Thus $g(X)$ is monic, has α as a zero and has the same degree n as $f(X)$. All of the coefficients of $g(X)$ are in $\mathbb{F}$ since $\mathbb{F}$ is a field. ■

Exercises 2.2

1. If α is a zero of $3X^3 - 2X + 1$, find a monic polynomial with coefficients in $\mathbb{Q}$ having α as a zero.
2. If $2\alpha^3 - 1 = \sqrt{3}$, find a monic polynomial in $\mathbb{Q}[X]$ which has α as a zero.
3. (a) Write down three different monic polynomials in $\mathbb{Q}[X]$ which have $\sqrt[3]{5}$ as a zero.
   (b) Write down three different monic polynomials in $\mathbb{Q}[X]$, each of degree 100, which have $\sqrt[3]{5}$ as a zero.
   (c) How many monic polynomials in $\mathbb{Q}[X]$ have $\sqrt[3]{5}$ as a zero?

2.3 Monic Polynomials of Least Degree

Even if we restrict attention to monic polynomials, there are still a lot of them which have $\sqrt{2}$ as a zero. For example, $\sqrt{2}$ is a zero of each of the polynomials

$$X^2 - 2, \quad X^4 - 4, \quad (X^2 - 2)^2, \quad (X^2 - 2)(X^5 + 3),$$
$$(X^2 - 2)(X^{100} + 84X^3 + 73)$$

all of which are in $\mathbb{Q}[X]$. What distinguishes $X^2 - 2$ from the other polynomials is that it has the **least** degree.

If a number α is a zero of one monic polynomial then (as suggested by Exercises 2.2 #3 there are infinitely many monic polynomials of arbitrarily high degree having α as a zero. Although there is no maximum possible degree for such polynomials, there is always a least possible degree. (For example, if we know α is a zero of a monic polynomial of degree 12, we can be sure that the least possible degree is one of the finitely-many numbers $1, 2, 3, \ldots, 11, 12$.)

The following proposition shows that by focusing on the least possible degree, we choose a unique polynomial.

2.3.1 Proposition. *If $\alpha \in \mathbb{C}$ is algebraic over a field $\mathbb{F} \subseteq \mathbb{C}$ then among all monic polynomials $f(X) \in \mathbb{F}[X]$ with $f(\alpha) = 0$ there is a unique one of least degree.*

Proof. Since α is algebraic, there is (by Proposition 2.2.2) a monic polynomial $f(X) \in \mathbb{F}[X]$ with $f(\alpha) = 0$. Hence there is one such polynomial of least possible degree.

To prove there is only one such polynomial, suppose there are two distinct polynomials, $f_1(X)$ and $f_2(X)$, each having the least degree n. Thus

$$f_1(X) = a_0 + a_1 X + \cdots + a_{n-1} X^{n-1} + X^n$$
$$f_2(X) = b_0 + b_1 X + \cdots + b_{n-1} X^{n-1} + X^n.$$

Hence if we put $f(X) = f_1(X) - f_2(X)$ we get

$$f(X) = c_0 + c_1 X + \ldots + c_{n-1} X^{n-1}, \quad \text{where } c_i = a_i - b_i.$$

We want to prove that $f_1(X) = f_2(X)$. To do this we show that $f(X)$=0.

We shall suppose that $f(X) \neq 0$ and then produce a contradic tion.

Clearly $f(X) \in \mathbb{F}[X]$, $f(\alpha) = 0$ and either $f(X)$ is zero or $f(X)$ has degree $< n$. Suppose $f(X) \neq 0$. By Proposition 2.2.2, there is a monic polynomial which has the same degree as $f(X)$ and which has α as a zero. But this is impossible since n is the least degree for which there is a monic polynomial having α as a zero. So we have a contradiction, and our supposition is false.

It follows that $f_1(X) = f_2(X)$. ■

The above theorem enables us to assign to each algebraic number a unique polynomial. This polynomial is so important that we introduce some terminology and notation for it.

2.3.2 Definition. Let $\alpha \in \mathbb{C}$ be algebraic over a field $\mathbb{F} \subseteq \mathbb{C}$. The unique polynomial of least degree among those polynomials $f(X)$ in $\mathbb{F}[X]$ satisfying

$$\text{(i) } f(\alpha) = 0 \quad \text{and} \quad \text{(ii) } f(X) \text{ is monic}$$

is called *the irreducible polynomial of α over* $\mathbb{F}$. This polynomial is denoted by

$$\operatorname{irr}(\alpha, \mathbb{F}).$$

Its degree is called the *degree of α over* $\mathbb{F}$ and is denoted by

$$\deg(\alpha, \mathbb{F}). \qquad \blacksquare$$

The word "irreducible" means "cannot be reduced" and its use in the above definition stems from the idea that we cannot reduce the degree below that of $\operatorname{irr}(\alpha, \mathbb{F})$ if we want to have a polynomial with the desired properties (i) and (ii).

Later you will find that $\operatorname{irr}(\alpha, \mathbb{F})$ is also irreducible in a different sense. Thus the use of the word "irreducible" is doubly justified.

2.3.3 Example. We prove that the irreducible polynomial of $\sqrt{2}$ over $\mathbb{Q}$ is $X^2 - 2$.

Proof. It is clear that this polynomial has $\sqrt{2}$ as a zero, that its coefficients are in $\mathbb{Q}$ and that it is monic. It remains to show there is no polynomial of smaller degree with these properties.

Suppose there is a polynomial of smaller degree with these properties. This polynomial must then be $\quad a + X, \quad$ for some $a \in \mathbb{Q}$.

But then we would have

$$a + \sqrt{2} = 0$$

so that $\sqrt{2} = -a \in \mathbb{Q}$, which is a contradiction.

Hence

$$\operatorname{irr}(\sqrt{2}, \mathbb{Q}) = X^2 - 2$$

and therefore

$$\deg(\sqrt{2}, \mathbb{Q}) = 2. \qquad \blacksquare$$

Note that, although the irreducible polynomial of $\sqrt{2}$ over $\mathbb{Q}$ is $X^2 - 2$, its irreducible polynomial over $\mathbb{R}$ is $X - \sqrt{2}$. This is because $X - \sqrt{2}$ is in $\mathbb{R}[X]$ but not in $\mathbb{Q}[X]$.

The idea in the worked example above is quite simple – find the polynomial you think is of least degree and, to show it really is of least degree, consider all smaller degrees. This can be far from straight-forward if the polynomial in question has degree greater than 2. For example, you would suspect that $\operatorname{irr}(\sqrt[7]{2}, \mathbb{Q})$ is $X^7 - 2$. But to prove that $\sqrt[7]{2}$ is not a zero of any monic polynomial in $\mathbb{Q}[X]$ of degrees 1, 2,

3, 4, 5 or 6 would be very difficult (to say the least) with the techniques you know at present. Special techniques have been developed for these sorts of problems; you will meet some of them in Chapter 4.

Exercises 2.3

1. (a) State the irreducible polynomial of $\sqrt{5}$ over $\mathbb{Q}$ and then prove your answer is correct.
   (b) What is $\deg(\sqrt{5}, \mathbb{Q})$?

2. In each case write down a nonzero polynomial $f(X)$ satisfying the stated conditions:

   (a) $f(X)$ is a monic polynomial over $\mathbb{Q}$ with $\sqrt{2}$ as a zero.
   (b) $f(X)$ is a polynomial over $\mathbb{Q}$ with $\sqrt{2}$ as a zero but is not monic.
   (c) $f(X)$ is a polynomial of least degree such that

   $$f(X) \in \mathbb{R}[X] \quad \text{and} \quad f(\sqrt{2}) = 0.$$

   (d) $f(X)$ is another polynomial satisfying conditions (c).
   (e) $f(X) \in \mathbb{Q}[X]$ and has both $\sqrt{2}$ and $\sqrt{3}$ as zeros.

3. (a) In each case write down the unique monic polynomial $f(X)$ of least degree which has the stated property:
   (i) $f(X) \in \mathbb{R}[X]$ and has $\sqrt{3}$ as a zero.
   (ii) $f(X) \in \mathbb{Q}[X]$ and has $\sqrt{3}$ as a zero.
   (b) Explain why the polynomials you gave in part (a) have least degree.
   (c) What do your answers to part (a) tell you about
   (i) $\operatorname{irr}(\sqrt{3}, \mathbb{R})$ and $\deg(\sqrt{3}, \mathbb{R})$?
   (ii) $\operatorname{irr}(\sqrt{3}, \mathbb{Q})$ and $\deg(\sqrt{3}, \mathbb{Q})$?

4. Assume α is algebraic over $\mathbb{F}$. Let n be the least degree of all monic nonzero polynomials $f(X)$ in $\mathbb{F}[X]$ with $f(\alpha) = 0$ and let m be the least degree of all nonzero polynomials $g(X)$ (monic or not) in $\mathbb{F}[X]$ with $g(\alpha) = 0$. Explain why $m = n$. (Which theorem or proposition from an earlier section do you use?)

5. Let $\mathbb{F}$ be a subfield of $\mathbb{R}$ and let $a, b, c \in \mathbb{F}$ with c positive.
   (a) Show that $a + b\sqrt{c}$ is a zero of a monic quadratic polynomial in $\mathbb{F}[X]$.
   (b) What can be said about the degree of $a + b\sqrt{c}$ over $\mathbb{F}$?

6. In accordance with #4, 5, 7 of Exercises 2.1, if $\alpha \in \mathbb{R}$ is algebraic over $\mathbb{F}$ then so are $-\alpha$, $\sqrt{\alpha}$ when $\alpha > 0$, and $1/\alpha$ when $\alpha \neq 0$.

   (a) What is the relationship between $\deg(\alpha, \mathbb{F})$ and $\deg(-\alpha, \mathbb{F})$?
   (b) Prove that $\deg(\sqrt{\alpha}, \mathbb{F}) \leq 2\deg(\alpha, \mathbb{F})$.
   (c) Is the following statement true or false? (Justify your answer.) If $\alpha \neq 0$ is algebraic over $\mathbb{F}$ then

$$\deg(\alpha, \mathbb{F}) = \deg\left(\frac{1}{\alpha}, \mathbb{F}\right).$$

7. Let $\mathbb{F} \subseteq \mathbb{C}$ and let $\alpha \in \mathbb{C}$. In each case decide whether the statement is true or false (and justify your answer).

   (a) If $\deg(\alpha, \mathbb{F}) = 2$ then $\alpha \notin \mathbb{F}$.
   (b) If $\deg(\alpha, \mathbb{F}) = 2$ then $\alpha^2 \in \mathbb{F}$.

8. If $\alpha \in \mathbb{C}$ is algebraic over a subfield $\mathbb{F}$ of $\mathbb{C}$ and

$$\{1, \alpha, \alpha^2, \alpha^3\}$$

   is linearly independent over $\mathbb{F}$, what can be said about $\deg(\alpha, \mathbb{F})$?

9. Let $\mathbb{F}$ be a subfield of $\mathbb{C}$ and let α be a complex number. Assume that the set $\{1, \alpha, \alpha^2\alpha^3, \alpha^4\}$ is linearly dependent over $\mathbb{F}$.

   (a) Prove that α is algebraic over $\mathbb{F}$.
   (b) What can be said about $\deg(\alpha, \mathbb{F})$? (Justify your answer.)

10. Let $\mathbb{F}$ be a subfield of $\mathbb{C}$ and let $\alpha \in \mathbb{C}$. Prove that $\deg(\alpha, F) = 1$ if and only if $\alpha \in \mathbb{F}$.

11. Is the following statement true or false? (Justify your answer.) If α, β and $\alpha\beta$ are algebraic over $\mathbb{Q}$ then

$$\deg(\alpha\beta, \mathbb{Q}) = \deg(\alpha, \mathbb{Q}) \deg(\beta, \mathbb{Q}).$$

12. Let $\mathbb{E} \subseteq \mathbb{F}$ be subfields of $\mathbb{C}$ and assume that $\beta \in \mathbb{C}$ is algebraic over $\mathbb{E}$.

    (a) Prove that β is algebraic over $\mathbb{F}$ also.
    (b) Prove that $\deg(\beta, \mathbb{E}) \geq \deg(\beta, \mathbb{F})$.

13. Let $\mathbb{E} \subseteq \mathbb{F}$ be subfields of $\mathbb{C}$ such that $[\mathbb{F} : \mathbb{E}] = 5$. Let $\beta \in \mathbb{F}$.

    (a) What is the maximum number of elements which can be in a subset of $\mathbb{F}$ which is linearly independent over $\mathbb{E}$?
    (b) State why the following powers of β are all in $\mathbb{F}$:

$$1, \beta, \beta^2, \beta^3, \beta^4, \beta^5.$$

    (c) Deduce from (a) and (b) that β is algebraic over $\mathbb{E}$.
    (d) What can be said about (i) $\deg(\beta, \mathbb{E})$? (ii) $\deg(\beta, \mathbb{F})$?

14. Let $\mathbb{F}$ be a subfield of $\mathbb{C}$ and let $\alpha \in \mathbb{C}$. Let $f(X) \in \mathbb{F}[X]$ be a monic polynomial with $f(\alpha) = 0$. If there are polynomials $g(X)$, $h(X)$ in $\mathbb{F}[X]$, each of degree ≥ 1, such that $f(X) = g(X)h(X)$, show that $f(X)$ is not $\operatorname{irr}(\alpha, \mathbb{F})$.

Additional Reading for Chapter 2

We have introduced $\text{irr}(\alpha, \mathbb{F})$ with very few preliminary results about polynomials. However, treatments of $\text{irr}(\alpha, \mathbb{F})$ in most other books depend heavily on techniques for showing that a particular polynomial cannot be reduced (that is, factorized further). For this reason, you may find it confusing to read about $\text{irr}(\alpha, \mathbb{F})$ in other books before you have finished reading our Chapter 4.

A discussion of numbers algebraic over a field is given in the book (Birkhoff and MacLane, 1953, §1 & §2 of Chapter XIV). For some historical facts about transcendental numbers, see Section 5.4 of Niven (1961). The delightful book Ribenboim (2000) provides a good coverage of a wide range of topics in Number Theory, and is written for the non-specialist.

In Exercises 2.1 #8 we outlined a proof that π is an irrational number. The book Berggren et al. (1997) provides a collection of significant articles written over hundreds of years on the number π. This is a remarkably good resource where you can read original articles, not easily accessible elsewhere.

Chapter 3
Extending Fields

If $\mathbb{F}$ is a subfield of $\mathbb{C}$ and α is a complex *number* which is algebraic over $\mathbb{F}$, we show how to construct a certain vector space $\mathbb{F}(\alpha)$ which contains α and which satisfies

$$\mathbb{F} \subseteq \mathbb{F}(\alpha) \subseteq \mathbb{C}.$$

This vector space is then shown to be a subfield of $\mathbb{C}$. Thus, from the field $\mathbb{F}$ and the number α, we have produced a larger field $\mathbb{F}(\alpha)$.

Fields of the form $\mathbb{F}(\alpha)$ are essential to our analysis of the lengths of those line segments which can be constructed with straightedge and compass.

3.1 An Illustration: $\mathbb{Q}(\sqrt{2})$

As $\mathbb{Q}$ is a subfield of $\mathbb{C}$, we can consider $\mathbb{C}$ as a vector space over $\mathbb{Q}$, taking the elements of $\mathbb{C}$ as the vectors and the elements of $\mathbb{Q}$ as the scalars.

3.1.1 Definition. The set $\mathbb{Q}(\sqrt{2}) \subseteq \mathbb{C}$ is defined by putting

$$\mathbb{Q}(\sqrt{2}) = \{a + b\sqrt{2} \colon a, b \in \mathbb{Q}\}.$$ ■

Thus $\mathbb{Q}(\sqrt{2})$ is the *linear span* of the set of vectors $\{1, \sqrt{2}\}$ over $\mathbb{Q}$ and is therefore a vector *subspace* of $\mathbb{C}$ over $\mathbb{Q}$. Hence $\mathbb{Q}(\sqrt{2})$ is a vector space over $\mathbb{Q}$.

3.1.2 Proposition. *The set of vectors* $\{1, \sqrt{2}\}$ *is a basis for the vector space* $\mathbb{Q}(\sqrt{2})$ *over* $\mathbb{Q}$.

Proof. As noted above, this set of vectors spans $\mathbb{Q}(\sqrt{2})$.

Hence it is sufficient to establish its linear independence over $\mathbb{Q}$. To this end, let $a, b \in \mathbb{Q}$ with

S. A. Morris et al., *Abstract Algebra and Famous Impossibilities*, Undergraduate Texts in Mathematics, https://doi.org/10.1007/978-3-031-05698-7_3

$$a + b\sqrt{2} = 0. \tag{1}$$

If $b \neq 0$, then $\sqrt{2} = -a/b$, which is again in $\mathbb{Q}$ as $\mathbb{Q}$ is a field. This contradicts the fact that $\sqrt{2}$ is irrational. Hence $b = 0$. It now follows from (1) that also $a = 0$.

Thus (1) implies $a = 0$ and $b = 0$ so that the set of vectors $\{1, \sqrt{2}\}$ is linearly independent over $\mathbb{Q}$. ■

Note that the dimension of the vector space $\mathbb{Q}(\sqrt{2})$ over $\mathbb{Q}$ is 2, this being the number of vectors in the basis

3.1.3 Proposition. *$\mathbb{Q}(\sqrt{2})$ is a ring. Indeed it is a subring of $\mathbb{C}$.*

Proof. To show that $\mathbb{Q}(\sqrt{2})$ is a subring of $\mathbb{C}$, it is sufficient to check that it is closed under addition and multiplication. The former is obvious while the latter is left as a simple exercise. ■

3.1.4 Proposition. *$\mathbb{Q}(\sqrt{2})$ is a field. Indeed it is a subfield of $\mathbb{C}$.*

Proof. To show that this subring of $\mathbb{C}$ is a subfield, it is sufficient to check that it contains the reciprocal of each of its nonzero elements.

So let $x \in \mathbb{Q}(\sqrt{2})$ be such that $x \neq 0$. Thus

$$x = a + b\sqrt{2}$$

where $a, b \in \mathbb{Q}$ and $a \neq 0$ or $b \neq 0$. It follows that

$$a - b\sqrt{2} \neq 0$$

by the linear independence of the set $\{1, \sqrt{2}\}$. Thus

$$\begin{aligned} \frac{1}{x} &= \frac{1}{a + b\sqrt{2}} \\ &= \frac{1}{a + b\sqrt{2}} \cdot \frac{a - b\sqrt{2}}{a - b\sqrt{2}} \\ &= \left(\frac{a}{a^2 - 2b^2}\right) + \left(\frac{-b}{a^2 - 2b^2}\right)\sqrt{2} \end{aligned}$$

which is again an element of $\mathbb{Q}(\sqrt{2})$, since $a/(a^2 - 2b^2)$ and $-b/(a^2 - 2b^2)$ are both in $\mathbb{Q}$. ■

The following proposition gives a way of describing $\mathbb{Q}(\sqrt{2})$ as a field with a certain property. The proof, like that of Proposition 1.1.1, involves proving a set inclusion.

3.1.5 Proposition. *$\mathbb{Q}(\sqrt{2})$ is the smallest field containing all the numbers in the field $\mathbb{Q}$ and the number $\sqrt{2}$.*

Proof. Let $\mathbb{F}$ be any field containing $\mathbb{Q}$ and $\sqrt{2}$.

> It is clear from what has been said earlier in this section that $\mathbb{Q}(\sqrt{2})$ is a field which contains both $\mathbb{Q}$ and $\sqrt{2}$.
>
> To show $\mathbb{Q}(\sqrt{2})$ is the smallest field containing both $\mathbb{Q}$ and $\sqrt{2}$, we shall prove that
>
> $$\mathbb{Q}(\sqrt{2}) \subseteq \mathbb{F}.$$

To prove that $\mathbb{Q}(\sqrt{2}) \subseteq \mathbb{F}$ we shall show that

$$\text{if } x \in \mathbb{Q}(\sqrt{2}), \quad \text{then } x \in \mathbb{F}. \tag{2}$$

To prove this we start by assuming the hypothesis of (2).
Let $x \in \mathbb{Q}(\sqrt{2})$; that is,

$$x = a + b\sqrt{2} \quad \text{for some } a, b \in \mathbb{Q}.$$

> Our aim is to prove that $x \in \mathbb{F}$.
>
> To do this we shall use the fact that $\mathbb{F}$ is a field. In particular, we shall use the fact that every field is closed under multiplication and addition.

By assumption, $\sqrt{2} \in \mathbb{F}$.
Also $a, b \in \mathbb{F}$ as $\mathbb{F}$ is assumed to contain $\mathbb{Q}$.
Hence $b\sqrt{2} \in \mathbb{F}$ as $\mathbb{F}$ is closed under multiplication.
So $a + b\sqrt{2} \in \mathbb{F}$ as $\mathbb{F}$ is closed under addition.
Thus $x \in \mathbb{F}$, as required. ∎

Because $\mathbb{Q}(\sqrt{2})$ is a field, the theory developed in Chapter 2 for a field $\mathbb{F} \subseteq \mathbb{C}$ can now be applied in the case $\mathbb{F} = \mathbb{Q}(\sqrt{2})$. The following example illustrates this.

3.1.6 Example. The number $\sqrt{3}$ is algebraic over $\mathbb{Q}(\sqrt{2})$ and has degree 2 over this field.

Proof. The polynomial $X^2 - 3$ has coefficients in $\mathbb{Q}$, and hence also in $\mathbb{Q}(\sqrt{2})$. It is monic, moreover, and has $\sqrt{3}$ as a zero. Hence $\sqrt{3}$ is algebraic over $\mathbb{Q}(\sqrt{2})$.

Suppose $X^2 - 3$ is not the irreducible polynomial of $\sqrt{3}$ over $\mathbb{Q}(\sqrt{2})$. Then the irreducible polynomial is a monic first degree polynomial

$$a + X$$

for some $a \in \mathbb{Q}(\sqrt{2})$. Hence $a + \sqrt{3} = 0$ and so $\sqrt{3} = -a$, which implies that

$$\sqrt{3} = c + d\sqrt{2}$$

for some $c, d \in \mathbb{Q}$. Squaring both sides gives

$$3 = c^2 + 2\sqrt{2}cd + 2d^2.$$

If

$$c = 0$$

we have $2d^2 = 3$ which leads to the contradiction that $\sqrt{3/2}$ is rational, while if

$$d = 0$$

we get the contradiction that $\sqrt{3}$ is rational. Hence $cd \neq 0$, which gives

$$\sqrt{2} = \frac{3 - c^2 - 2d^2}{2cd} \in \mathbb{Q},$$

which is also a contradiction.

Thus our original supposition that $X^2 - 3$ is not the irreducible polynomial of $\sqrt{3}$ over $\mathbb{Q}(\sqrt{2})$ has led to a contradiction. Hence $X^2 - 3$ is the irreducible polynomial of $\sqrt{3}$ over $\mathbb{Q}(\sqrt{2})$; that is,

$$\operatorname{irr}(\sqrt{3},\ \mathbb{Q}(\sqrt{2})) = X^2 - 3$$

and hence

$$\deg(\sqrt{3},\ \mathbb{Q}(\sqrt{2})) = 2.$$

■

Exercises 3.1

1. Verify that each of the following numbers is in $\mathbb{Q}(\sqrt{2})$ by expressing it in the form $a + b\sqrt{2}$, where $a, b \in \mathbb{Q}$.
   (i) $(2 + 3\sqrt{2})(1 - 2\sqrt{2})$,
   (ii) $(1 + 3\sqrt{2})/(3 + 2\sqrt{2})$,
   (iii) $\sqrt{3 + 2\sqrt{2}}$.
   [Hint. Let this equal $a + b\sqrt{2}$ and square both sides.]

2. Show that the number $\sqrt{1+\sqrt{2}}$ is not in $\mathbb{Q}(\sqrt{2})$.
[Hint. Suppose this number equals $a+b\sqrt{2}$, square both sides and derive a contradiction.]

3. Show that the zeros in $\mathbb{C}$ of the polynomial X^2+2X-7 are both in $\mathbb{Q}(\sqrt{2})$.

4. Verify that the number $\sqrt{5}$ is algebraic over $\mathbb{Q}(\sqrt{2})$ and find its degree over this field.

5. Verify that the number $\sqrt{144}$ is algebraic over $\mathbb{Q}(\sqrt{2})$ and find its degree over this field.

6. Complete the proof of Proposition 3.1.3, begun in the text.

   [Hint. Assume that $x, y \in \mathbb{Q}(\sqrt{2})$ and then spell out explicitly what this means. Hence show that $xy \in \mathbb{Q}(\sqrt{2})$.]

7. Verify that $\sqrt{m}$ is algebraic over each of the fields $\mathbb{Q}$ and $\mathbb{Q}(\sqrt{2})$, for each integer $m \geq 0$. State in which cases it is true that

$$\deg(\sqrt{m},\ \mathbb{Q}) = \deg(\sqrt{m},\ \mathbb{Q}(\sqrt{2})).$$

8. Let $a, b \in \mathbb{Q}$. Show that if $a+b\sqrt{2}$ is a zero of a quadratic polynomial in $\mathbb{Q}[X]$ then its "conjugate" $a-b\sqrt{2}$ is also a zero of this polynomial.

9.* Prove the result of Exercise 8, but without the restriction that the polynomial is quadratic.

3.2 Construction of $\mathbb{F}(\alpha)$

In the previous section we constructed, from the field $\mathbb{Q}$ and the number $\sqrt{2}$, a vector subspace $\mathbb{Q}(\sqrt{2})$ of $\mathbb{C}$ by putting

$$\mathbb{Q}(\sqrt{2}) = \{a+b\sqrt{2} \colon a,\ b \in \mathbb{Q}\}.$$

We then found that $\mathbb{Q}(\sqrt{2})$ is closed under multiplication (as well as addition) and hence is a ring. Next $\mathbb{Q}(\sqrt{2})$ was shown to contain the reciprocal of each of its nonzero elements, and hence it is also a field. Finally we showed that $\mathbb{Q}(\sqrt{2})$ was the *smallest subfield of* $\mathbb{C}$ *containing all the numbers in* $\mathbb{Q}$ *together with the number* $\sqrt{2}$.

In this section we generalize the construction of $\mathbb{Q}(\sqrt{2})$ by constructing a subset $\mathbb{F}(\alpha)$ of $\mathbb{C}$ for any subfield $\mathbb{F}$ of $\mathbb{C}$ and any number $\alpha \in \mathbb{C}$ which is algebraic over $\mathbb{F}$. We begin by defining $\mathbb{F}(\alpha)$ and then showing that this set is in turn a vector subspace, a subring, and then a subfield of $\mathbb{C}$. Our final goal is Theorem 3.2.8, which says that $\mathbb{F}(\alpha)$ is *the smallest subfield of* $\mathbb{C}$ *containing* α *and all the numbers in* $\mathbb{F}$.

Throughout this section, we assume that α is algebraic over $\mathbb{F}$.

3.2.1 Definition. Let $\mathbb{F}$ be a subfield of $\mathbb{C}$ and let $\alpha \in \mathbb{C}$ be algebraic over $\mathbb{F}$ with $\deg(\alpha, \mathbb{F}) = n$. The *extension of* $\mathbb{F}$ *by* α is the set $\mathbb{F}(\alpha) \subseteq \mathbb{C}$ where

$$\mathbb{F}(\alpha) = \{b_0 + b_1\alpha + \ldots + b_{n-1}\alpha^{n-1} \ : \ b_0, \ b_1, \ldots, \ b_{n-1} \in \mathbb{F}\}. \qquad \blacksquare$$

Thus $\mathbb{F}(\alpha)$ is the linear span over $\mathbb{F}$ of the powers

$$1, \ \alpha, \ \alpha^2, \ldots, \alpha^{n-1}$$

and so is a *vector subspace* of $\mathbb{C}$ over $\mathbb{F}$. In this section we show that $\mathbb{F}(\alpha)$ is a field – indeed it is the smallest field containing $\mathbb{F}$ and α. These results are proved in Theorems 3.2.6 and 3.2.8.

Why did we stop at the power α^{n-1}? The answer is given by the following result, which shows that any further powers would be redundant.

3.2.2 Proposition. *The set* $\mathbb{F}(\alpha)$ *contains all the remaining positive powers of* α*:*

$$\alpha^n, \ \alpha^{n+1}, \alpha^{n+2}, \ \alpha^{n+3}, \ldots.$$

Proof. Since $n = \deg(\alpha, \mathbb{F})$, α is a zero of a monic polynomial of degree n in $\mathbb{F}[X]$. Hence

$$c_0 + c_1\alpha + \ldots + c_{n-1}\alpha^{n-1} + \alpha^n = 0$$

for some coefficients $c_0, c_1, \ldots, c_{n-1} \in \mathbb{F}$. Thus

$$\alpha^n = -c_0 - c_1\alpha - \ldots - c_{n-1}\alpha^{n-1} \tag{1}$$

and each $-c_i \in \mathbb{F}$, as $\mathbb{F}$ is a field. Hence, by the definition of $\mathbb{F}(\alpha)$ (Definition 3.2.1),

$$\alpha^n \in \mathbb{F}(\alpha).$$

If we multiply both sides of (1) by α we see that

$$\alpha^{n+1} = -c_0\alpha - \ldots - c_{n-2}\alpha^{n-1} - c_{n-1}\alpha^n. \tag{2}$$

Now $\alpha, \ldots, \alpha^{n-1}$ are all in $\mathbb{F}(\alpha)$ (by definition) and we have just shown that $\alpha^n \in \mathbb{F}(\alpha)$. Thus, because $\mathbb{F}(\alpha)$ is a vector subspace of $\mathbb{C}$, it follows from (2) that

$$\alpha^{n+1} \in \mathbb{F}(\alpha).$$

Now multiply both sides of (1) by α^2 and then proceed in the same way, thereby showing that

$$\alpha^{n+2} \in \mathbb{F}(\alpha).$$

It is clear that by proceeding in this way we can show that the powers

$$\alpha^n, \alpha^{n+1}, \alpha^{n+2}, \ldots$$

all belong to $\mathbb{F}(\alpha)$. ■

Thus, in our definition of $\mathbb{F}(\alpha)$ we have, in some sense, included "enough" powers of α. Have we included "too many"? An answer can be deduced from the following result.

3.2.3 Proposition. *If n is the degree of α over $\mathbb{F}$, then the set of vectors $\{1, \alpha, \alpha^2, \ldots, \alpha^{n-1}\}$ is linearly independent over $\mathbb{F}$.*

Proof. Suppose on the contrary that this set is linearly dependent; so there are scalars $c_0, c_1, \ldots, c_{n-1} \in \mathbb{F}$, not all zero, such that

$$c_0 + c_1\alpha + \ldots + c_{n-1}\alpha^{n-1} = 0.$$

Among the coefficients $c_0, c_1, \ldots, c_{n-1}$ pick the one, say c_i, farthest down the list which is nonzero. Dividing by this coefficient gives

$$\left(\frac{c_0}{c_i}\right) + \left(\frac{c_1}{c_i}\right)\alpha + \ldots + \left(\frac{c_{i-1}}{c_i}\right)\alpha^{i-1} + \alpha^i = 0$$

where $i \leq n - 1$, and the coefficients are still in the field $\mathbb{F}$.

Hence α is a zero of a monic polynomial in $\mathbb{F}[X]$ with degree smaller than n (which was the least possible degree for such polynomials). This contradiction shows that our supposition is false. ■

What if we tried to take

$$S = \{b_0 + b_1\alpha + \ldots + b_{n-2}\alpha^{n-2} : b_0, \ldots, b_{n-2} \in \mathbb{F}\}$$

as a candidate for a ring (or field) containing $\mathbb{F}$ and α. Then (if $n \geq 2$), α and α^{n-2} are both in S but their product α^{n-1} is not in S by the above proposition. Thus S is too small and so we need all the powers $1, \alpha, \ldots, \alpha^{n-1}$ in $\mathbb{F}(\alpha)$ for it to be a ring or a field.

3.2.4 Theorem. [Basis for $\mathbb{F}(\alpha)$ Theorem] Let $\mathbb{F}$ be a subfield of $\mathbb{C}$ and let $\alpha \in \mathbb{C}$ be algebraic over $\mathbb{F}$ with $deg(\alpha, \mathbb{F}) = n$. Then the set of vectors $\{1, \alpha, \alpha^2, \ldots, \alpha^{n-1}\}$ is a basis for the vector space $\mathbb{F}(\alpha)$ over $\mathbb{F}$. In particular this vector space has dimension n, the degree of α over $\mathbb{F}$.

Proof. By Definition 3.2.1, this set of vectors spans the vector space $\mathbb{F}(\alpha)$ over $\mathbb{F}$ and, by Proposition 3.2.3, the set of vectors is linearly independent. Hence it is a basis for the vector space and the number n of elements in the basis is the dimension of the vector space. ■

As $\mathbb{F}(\alpha)$ is a subset of $\mathbb{C}$, its elements can be multiplied together. This leads to the following proposition.

3.2.5 Proposition. *Let* $\mathbb{F}$ *be a field,* $\alpha \in \mathbb{C}$ *and* α *algebraic of* $\mathbb{C}$ *over* $\mathbb{F}$*, then* $\mathbb{F}(\alpha)$ *is a ring. Indeed* $\mathbb{F}(\alpha)$ *is a subring of* $\mathbb{C}$.

Proof. To show $\mathbb{F}(\alpha)$ is a subring of $\mathbb{C}$, it is sufficient to show it is closed under addition and multiplication. The former is obvious while the latter will now be shown. Consider any two elements in $\mathbb{F}(\alpha)$, say

$$p = b_0 + b_1\alpha + b_2\alpha^2 + \ldots + b_{n-1}\alpha^{n-1} \in \mathbb{F}(\alpha)$$

and

$$q = c_0 + c_1\alpha + c_2\alpha^2 + \ldots + c_{n-1}\alpha^{n-1} \in \mathbb{F}(\alpha),$$

where the b's and c's belong to $\mathbb{F}$. If we multiply these two elements, we get a linear combination of powers of α, with coefficients still in the field $\mathbb{F}$. Because all positive powers of α are in $\mathbb{F}(\alpha)$ (by Proposition 3.2.2) and because $\mathbb{F}(\alpha)$ is closed under addition and scalar multiplication, it follows that the product pq is also in $\mathbb{F}(\alpha)$. ■

At long last we are in a position to establish that $\mathbb{F}(\alpha)$ is a field.

3.2.6 Theorem. *Let* $\mathbb{F}$ be a subfield of $\mathbb{C}$ and let $\alpha \in \mathbb{C}$ be algebraic over $\mathbb{F}$. Then $\mathbb{F}(\alpha)$ is a field. Indeed $\mathbb{F}(\alpha)$ is a subfield of $\mathbb{C}$.

Proof. In view of Proposition 3.2.5, what remains to be shown is that $1/\beta$ is in $\mathbb{F}(\alpha)$ for every nonzero β in $\mathbb{F}(\alpha)$. So let β be a nonzero number in $\mathbb{F}(\alpha)$.

Firstly note that $\{1, \beta, \beta^2, \ldots, \beta^n\}$ is a set of $n+1$ numbers which are all in $\mathbb{F}(\alpha)$, since $\mathbb{F}(\alpha)$ is closed under multiplication. Because $\mathbb{F}(\alpha)$ is an n–dimensional vector space over $\mathbb{F}$, this set must be linearly dependent, which means that there are scalars $d_0, d_1, \ldots, d_k$ in $\mathbb{F}$ (not all zero) with $k \leq n$ such that

$$d_0 + d_1\beta + d_2\beta^2 + \ldots + d_k\beta^k = 0. \tag{3}$$

If $d_0 = 0$ in (3), we could divide by β (or multiply by $1/\beta$) to reduce the number of terms in (3). By repeating this if necessary, we see that (3) can be assumed to hold with $d_0 \neq 0$. If we multiply (3) by $-1/d_0$ we have

$$-1 + e_1\beta + e_2\beta^2 + \ldots + e_k\beta^k = 0,$$

where each $e_i = -d_i/d_0$ is in $\mathbb{F}$ since $\mathbb{F}$ is a field. Thus

$$1 = e_1\beta + e_2\beta^2 + \ldots + e_k\beta^k = \beta(e_1 + e_2\beta + \ldots + e_k\beta^{k-1}).$$

It follows that

$$1/\beta = e_1 + e_2\beta + \ldots + e_k\beta^{k-1},$$

which is in $\mathbb{F}(\alpha)$ since the e_i and β^i are in $\mathbb{F}(\alpha)$ and $\mathbb{F}(\alpha)$ is a ring. ■

In simple cases the method used to prove Theorem 3.2.6 can be used to find a formula for $1/\beta$, as the following example shows.

3.2.7 Example. In $\mathbb{Q}(\sqrt{2})$, if $\beta = 1 + 2\sqrt{2}$ then $\beta^2 = 9 + 4\sqrt{2}$ and it is easy to see that $\beta^2 - 2\beta - 7 = 0$. This can be rewritten as $\beta(\beta - 2) = 7$ so that

$$1/\beta = (\beta - 2)/7 = -\frac{1}{7} + \frac{2}{7}\sqrt{2}.$$

■

3.2.8 Theorem. [Smallest Field Theorem] Let $\mathbb{F}$ be a subfield of $\mathbb{C}$ and let $\alpha \in \mathbb{C}$ be algebraic over $\mathbb{F}$. Then $\mathbb{F}(\alpha)$ is the smallest field containing α and all the numbers in $\mathbb{F}$.

Proof. This is analogous to the proof of Proposition 3.1.5. ■

The Smallest Field Theorem gives a way of describing $\mathbb{F}(\alpha)$ which is just as useful for many purposes as the definition of $\mathbb{F}(\alpha)$ (Definition 3.2.1). Hence, in solving problems involving $\mathbb{F}(\alpha)$ you should keep both Definition 3.2.1 and Theorem 3.2.8 in mind and then use whichever seems more appropriate for the particular problem.

In some books the rôles of our Smallest Field Theorem and our Definition 3.2.1 are reversed, the statement of the former appearing as a *definition* and the latter as a *theorem*. Our treatment is perhaps more concrete and so hopefully easier for those studying the topic for the first time.

Our approach assumes, however, that α is algebraic over $\mathbb{F}$. If this is not the case, our definition of $\mathbb{F}(\alpha)$ is not applicable. Hence, in such cases, we would use the statement of Theorem 3.2.8 as our definition.

The following example is a typical application of the Smallest Field Theorem.

3.2.9 Example. $\mathbb{F}(\alpha^2) \subseteq \mathbb{F}(\alpha)$ for each subfield $\mathbb{F}$ of $\mathbb{C}$ and each $\alpha \in \mathbb{C}$.

Proof. Let $\mathbb{F}$ be a subfield of $\mathbb{C}$ and let $\alpha \in \mathbb{C}$.
By the Smallest Field Theorem, $\mathbb{F}(\alpha)$ contains α and $\mathbb{F}$.
Hence $\mathbb{F}(\alpha)$ contains $\alpha.\alpha = \alpha^2$ since, being a field, $\mathbb{F}(\alpha)$ is closed under multiplication.
This shows that $\mathbb{F}(\alpha)$ is a field containing α^2 and $\mathbb{F}$.
But $\mathbb{F}(\alpha^2)$ is the smallest such field.
Therefore $\mathbb{F}(\alpha^2) \subseteq \mathbb{F}(\alpha)$, as required. ■

Exercises 3.2

1. What does the definition of $\mathbb{F}(\alpha)$ say in each of the following cases?
   (i) $\deg(\alpha,\ \mathbb{F}) = 1$,
   (ii) $\deg(\alpha,\ \mathbb{F}) = 2$,
   (iii) $\deg(\alpha,\ \mathbb{F}) = 3$.

2. (a) What does the definition of $\mathbb{F}(\alpha)$ tell you in the special case that $\mathbb{F} = \mathbb{Q}$ and $\alpha = \sqrt{3}$?
   (b) What do Propositions 3.2.2, 3.2.3 and Theorem 3.2.4 tell you in this special case?
   (c) Verify directly that
      (i) the set $\{1,\ \sqrt{3}\}$ is linearly independent over $\mathbb{Q}$,
      (ii) $\mathbb{Q}(\sqrt{3})$ contains the product $(2 + \sqrt{3})(3 + \sqrt{3})$ and the quotient $(2 + \sqrt{3})/(3 + \sqrt{3})$.

3. What is the degree of $\sqrt{2}$ over $\mathbb{R}$? Which familiar field is $\mathbb{R}(\sqrt{2})$?

4. What is the degree of i over $\mathbb{R}$? Which familiar field is $\mathbb{R}(i)$?

5. (a) State why $1 + \sqrt{2} \in \mathbb{Q}(\sqrt{2})$ and why $\sqrt{2} \in \mathbb{Q}(1 + \sqrt{2})$.
   (b) Prove that $\mathbb{Q}(1 + \sqrt{2}) = \mathbb{Q}(\sqrt{2})$ by showing that each field is a subset of the other. (You may use the Smallest Field Theorem.)

6. (a) Write down the irreducible polynomial of $\sqrt{3}$ over each of the following fields:
   (i) $\mathbb{Q}$, (ii) $\mathbb{R}$, (iii) $\mathbb{Q}(\sqrt{3})$.
   (b) State the degree of $\sqrt{3}$ over each of these fields.

7. Find $\mathrm{irr}(\sqrt{2},\ \mathbb{Q}(1 + \sqrt{2}))$ and $\deg(\sqrt{2},\ \mathbb{Q}(1 + \sqrt{2}))$. (Justify your answer.)

8. (a) Prove that $\sqrt{2} \notin \mathbb{Q}(\sqrt{3})$.
   (b) Show that $\sqrt{2}$ is algebraic over $\mathbb{Q}(\sqrt{3})$.
   (c) Write down $\mathrm{irr}(\sqrt{2},\ \mathbb{Q}(\sqrt{3}))$.
   (d) Justify your answer to (c). (Use (a).)

9. (a) What does your answer to Exercise 8(c) tell you about $\deg(\sqrt{2},\ \mathbb{Q}(\sqrt{3}))$?
   (b) Let $\mathbb{F}$ denote $\mathbb{Q}(\sqrt{3})$. Use (a) to write down a basis for the vector space $\mathbb{F}(\sqrt{2})$ over $\mathbb{F}$.

10. Complete the proof of Proposition 3.2.5 by showing that $\mathbb{F}(\alpha)$ is closed under addition.

11. Let $\mathbb{F}$ be a subfield of $\mathbb{C}$ and let $\alpha,\ \beta \in \mathbb{C}$.
    (a) Prove that if $\mathbb{F}(\beta) \subseteq \mathbb{F}(\alpha)$ then $\beta \in \mathbb{F}(\alpha)$.
    (b) Prove the converse of the result stated in part (a).
    [Hint. Use the Smallest Field Theorem.]
    (c) Deduce that $\mathbb{F}(\alpha) = \mathbb{F}(\beta)$ if and only if $\alpha \in \mathbb{F}(\beta)$ and $\beta \in \mathbb{F}(\alpha)$.

12. Let $\beta = 1 + \sqrt[3]{2}$. Follow the method used in the proof of Theorem 3.2.6 and in Example 3.2.7 to obtain a formula in $\mathbb{Q}(\sqrt[3]{2})$ for $1/\beta$.

13. Let $\mathbb{F}$ be a subfield of $\mathbb{C}$ and let α be a nonzero complex number. Show that if $1/\alpha$ can be expressed in the form

$$d_0 + d_1\alpha + \ldots + d_k\alpha^k$$

for $k \geq 1$ and $d_i \in \mathbb{F}$, then α is algebraic over $\mathbb{F}$.
[Remark. It follows that if $\mathbb{F}(\alpha)$ is defined to be the linear span of all positive powers $1, \alpha, \alpha^2, \ldots$ of α (much as in Definition 3.2.1) then $\mathbb{F}(\alpha)$ is a field precisely when α is algebraic over $\mathbb{F}$.]

14.* Find all the fields $\mathbb{F}$ such that $\mathbb{F}(\sqrt{2}) = \mathbb{Q}(\sqrt{2})$. (Justify your answer.)

15.* Describe the field $\mathbb{Q}(\pi)$ as explicitly as possible.
[Note that π is not algebraic over $\mathbb{Q}$ and so Definition 3.2.1 is not applicable. Instead, $\mathbb{Q}(\pi)$ is defined as the smallest subfield of $\mathbb{C}$ which contains $\mathbb{Q}$ and π.]

3.3 Iterating the Construction

The starting point for the previous section was a field $\mathbb{F} \subseteq \mathbb{C}$ and a number $\alpha \in \mathbb{C}$ which was algebraic over $\mathbb{F}$. From these ingredients a new field was constructed,

$$\mathbb{F}(\alpha) \subseteq \mathbb{C}.$$

We can now take the field $\mathbb{F}(\alpha)$ and a number $\beta \in \mathbb{C}$ which is algebraic over $\mathbb{F}(\alpha)$ as the starting point for a further application of the construction process to get a further new field

$$\mathbb{F}(\alpha)(\beta) \subseteq \mathbb{C}.$$

This process can be repeated as often as we like to give a "tower" of fields, each inside the next,

$$\mathbb{Q} \subseteq \mathbb{F} \subseteq \mathbb{F}(\alpha) \subseteq \mathbb{F}(\alpha)(\beta) \subseteq \mathbb{F}(\alpha)(\beta)(\gamma) \subseteq \ldots \subseteq \mathbb{C}.$$

3.3.1 Example. $\mathbb{Q}(\sqrt{2})(\sqrt{3})$ is the linear span over $\mathbb{Q}(\sqrt{2})$ of the set of vectors $\{1, \sqrt{3}\}$. This set, furthermore, is a basis for the vector space $\mathbb{Q}(\sqrt{2})(\sqrt{3})$ over $\mathbb{Q}(\sqrt{2})$.

Proof. Firstly note that by the solution to Example 3.1.6,

$$\mathrm{irr}(\sqrt{3},\ \mathbb{Q}(\sqrt{2})) = \mathrm{X}^2 - 3$$

and hence

$$\deg(\sqrt{3},\ \mathbb{Q}(\sqrt{2})) = 2.$$

It now follows from Definition 3.2.1 that

$$\mathbb{Q}(\sqrt{2})(\sqrt{3}) = \{x + \sqrt{3}y : x,\ y \in \mathbb{Q}(\sqrt{2})\},$$

which is just the linear span of the set of vectors $\{1, \sqrt{3}\}$ over $\mathbb{Q}(\sqrt{2})$. That this set of vectors forms a basis follows from Theorem 3.2.4. ■

The tower of fields

$$\mathbb{Q} \subseteq \mathbb{Q}(\sqrt{2}) \subseteq \mathbb{Q}(\sqrt{2})(\sqrt{3})$$

invites us to consider $\mathbb{Q}(\sqrt{2})(\sqrt{3})$ as a vector space over $\mathbb{Q}$ also. This leads to the following example.

3.3.2 Example. $\mathbb{Q}(\sqrt{2})(\sqrt{3})$ is the linear span over $\mathbb{Q}$ of the set of vectors $\{1,\ \sqrt{2},\ \sqrt{3},\ \sqrt{2}\sqrt{3}\}$.

Proof. By the previous example

$$\begin{aligned}\mathbb{Q}(\sqrt{2})(\sqrt{3}) &= \{x + \sqrt{3}y : x, y \in \mathbb{Q}(\sqrt{2})\}\\ &= \{(a + b\sqrt{2}) + \sqrt{3}(c + d\sqrt{2}) : a, b, c, d \in \mathbb{Q}\}\\ &= \{a + b\sqrt{2} + c\sqrt{3} + d\sqrt{2}\sqrt{3} : a, b, c, d \in \mathbb{Q}\}\end{aligned}$$

which expresses it as the required linear span. ■

One might guess from the above example that the set of vectors $\{1,\ \sqrt{2},\ \sqrt{3},\ \sqrt{2}\sqrt{3}\}$ is in fact a basis for the vector space $\mathbb{Q}(\sqrt{2})(\sqrt{3})$ over $\mathbb{Q}$. One of the aims of the next section is to prove a theorem from which this result will follow very easily.

Exercises 3.3

1. Find a spanning set for $\mathbb{Q}(\sqrt{2})(i)$
   (a) as a vector space over $\mathbb{Q}(\sqrt{2})$,
   (b) as a vector space over $\mathbb{Q}$.
2. Simplify $\mathbb{F}(\alpha)(1 + \alpha)$, where $\mathbb{F}$ is a subfield of $\mathbb{C}$ and $\alpha \in \mathbb{C}$. (Justify your answer.)
3. Write down three different extension fields of $\mathbb{Q}$ which contain $\sqrt{2}$.
4. If $\mathbb{F}$ is a subfield of $\mathbb{C}$ and $\alpha, \beta \in \mathbb{C}$ then $\mathbb{F}(\alpha,\ \beta)$ is defined to be the smallest subfield of $\mathbb{C}$ which contains all numbers in $\mathbb{F}$ and also the numbers α and β. Use this definition to show that, if α and β are algebraic over $\mathbb{F}$, then

$$\mathbb{F}(\alpha)(\beta) = \mathbb{F}(\alpha,\ \beta) = \mathbb{F}(\beta)(\alpha)$$

where, as in this section, $\mathbb{F}(\alpha)(\beta)$ means the smallest field containing all numbers in the field $\mathbb{F}(\alpha)$ and the number β.

3.4 Towers of Fields

The aim of this section is to produce a theorem which is of vital importance in questions involving a tower

$$\mathbb{E} \subseteq \mathbb{F} \subseteq \mathbb{K}$$

consisting of three distinct subfields $\mathbb{E}$, $\mathbb{F}$ and $\mathbb{K}$ of $\mathbb{C}$. Implicit in this set-up are three different vector spaces and the theorem to be proved will give a precise relationship between their dimensions.

The three vector spaces arising from the tower are as follows. Since $\mathbb{E}$ is a subfield of $\mathbb{F}$, we may take $\mathbb{F}$ as the vectors and $\mathbb{E}$ as the scalars to give the vector space

(i) $\mathbb{F}$ over $\mathbb{E}$.

Likewise there are the vector spaces

(ii) $\mathbb{K}$ over $\mathbb{F}$, and
(iii) $\mathbb{K}$ over $\mathbb{E}$.

It may be helpful to see all these vector spaces together on a single diagram, as shown in Fig. 3.1

Here $\mathbb{F}$ plays a schizophrenic rôle: looking back we regard its elements as vectors, but looking forward we take them as scalars.

The theorem on dimensions will follow easily from the following theorem.

3.4.1 Theorem. [Basis for a Tower Theorem] Consider a tower of subfields of $\mathbb{C}$,

$$\mathbb{E} \subseteq \mathbb{F} \subseteq \mathbb{K}.$$

If the vector space $\mathbb{F}$ over $\mathbb{E}$ has a basis $\{\alpha_1, \ldots, \alpha_m\}$ and the vector space $\mathbb{K}$ over $\mathbb{F}$ has a basis $\{\beta_1, \ldots, \beta_n\}$, then the set of vectors

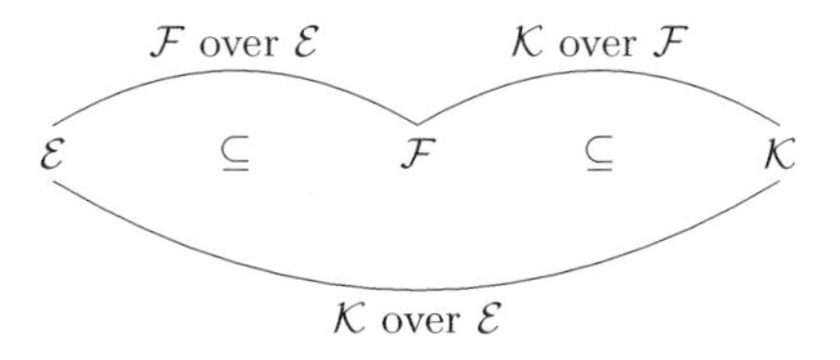

Fig. 3.1 Tower of fields

$$\begin{array}{cccc}\alpha_1\beta_1, & \alpha_1\beta_2, & \ldots, & \alpha_1\beta_n,\\ \alpha_2\beta_1, & \alpha_2\beta_2, & \ldots, & \alpha_2\beta_n,\\ \vdots & \vdots & & \vdots\\ \alpha_m\beta_1, & \alpha_m\beta_2, & \ldots, & \alpha_m\beta_n,\end{array}$$

forms a basis for the vector space $\mathbb{K}$ over $\mathbb{E}$.

Proof. The following diagram may help you to keep track of where the vectors come from.

$$\mathbb{E} \subseteq \underset{\alpha_j\in\mathbb{F}}{\mathbb{F}} \subseteq \underset{\substack{\alpha_j\beta_i\in\mathbb{K}\\ \beta_i\in\mathbb{K}}}{\mathbb{K}}$$

Firstly we show that the given set of vectors *spans* the vector space $\mathbb{K}$ over $\mathbb{E}$. To do this, let $k \in \mathbb{K}$.

Our aim is to prove that k is a linear combination of the $\alpha_j\beta_i$'s with coefficients in $\mathbb{E}$.

To do this we shall use the fact that the β's form a basis for $\mathbb{K}$ over $\mathbb{F}$ and the fact that the α's form a basis for $\mathbb{F}$ over $\mathbb{E}$.

Because the β's form a basis for $\mathbb{K}$ over $\mathbb{F}$, there exist n scalars $f_1, \ldots, f_n \in \mathbb{F}$ such that

$$k = \sum_{i=1}^{n} f_i\beta_i. \tag{1}$$

But since the α's form a basis for $\mathbb{F}$ over $\mathbb{E}$, there exist scalars $e_{ij}(1 \le i \le n, 1 \le j \le m)$ in $\mathbb{E}$ such that

$$f_i = \sum_{j=1}^{m} e_{ij}\alpha_j. \tag{2}$$

Substituting (2) in (1) gives

$$\begin{aligned} k &= \sum_{i=1}^{n}\left(\sum_{j=1}^{m} e_{ij}\alpha_j\right)\beta_i\\ &= \sum_{i=1}^{n}\sum_{j=1}^{m} e_{ij}(\alpha_j\beta_i). \end{aligned}$$

Thus the $\alpha_j\beta_i$'s span the vector space $\mathbb{K}$ over $\mathbb{E}$.

Secondly we show that these vectors are *linearly independent* over $\mathbb{E}$.

Let e_{ij} $(1 \le i \le n,\ 1 \le j \le m)$ be elements of $\mathbb{E}$ such that

$$\sum_{i=1}^{n}\sum_{j=1}^{m} e_{ij}\alpha_j\beta_i = 0. \tag{3}$$

Our aim is to prove that all the e's are zero.

To do this we shall examine (3) and use the fact that the β's are linearly independent over $\mathbb{F}$ and the fact that the α's are linearly independent over $\mathbb{E}$.

We can rewrite (3) as

$$\sum_{i=1}^{n}\left(\sum_{j=1}^{m} e_{ij}\alpha_j\right)\beta_i = 0.$$

But in this sum, the coefficients of the β_i's (each coefficient is a sum of the $e_{ij}\alpha_j$'s) are elements of $\mathbb{F}$ and the β's are linearly independent over $\mathbb{F}$. Hence, for $1 \le i \le n$,

$$\sum_{j=1}^{m} e_{ij}\alpha_j = 0.$$

But the α's are linearly independent over $\mathbb{E}$, which means that, for $1 \le j \le m$,

$$e_{ij} = 0.$$

This establishes the required linear independence.

Thus the $\alpha_j\beta_i$'s span the vector space $\mathbb{K}$ over $\mathbb{E}$ and are linearly independent over $\mathbb{E}$. Hence they form a basis for the vector space $\mathbb{K}$ over $\mathbb{E}$. ■

3.4.2 Example. The vector space $\mathbb{Q}(\sqrt{2})(\sqrt{3})$ over $\mathbb{Q}$ has the set of vectors

$$\{1,\ \sqrt{2},\ \sqrt{3},\ \sqrt{2}\sqrt{3}\}$$

as a basis.

Proof. Consider the tower

$$\mathbb{Q} \subseteq \mathbb{Q}(\sqrt{2}) \subseteq \mathbb{Q}(\sqrt{2})(\sqrt{3}).$$

By the Basis for $\mathbb{F}(\alpha)$ Theorem 3.2.4 and Example 2.3.3, the vector space $\mathbb{Q}(\sqrt{2})$ over $\mathbb{Q}$ has a basis $\{1, \sqrt{2}\}$ while by the same theorem and Example 3.1.6, the vector space $\mathbb{Q}(\sqrt{2})(\sqrt{3})$ over $\mathbb{Q}(\sqrt{2})$ has a basis $\{1, \sqrt{3}\}$. Hence, by the Basis for a Tower Theorem 3.4.1, the vector space $\mathbb{Q}(\sqrt{2})(\sqrt{3})$ over $\mathbb{Q}$ has the basis

$$\{1, \sqrt{2}, \sqrt{3}, \sqrt{2}\sqrt{3}\}.$$ ■

By counting the numbers of elements in the bases for the vector spaces occurring in Theorem 3.4.1 we can get a relationship between their dimensions. This gives us the following very useful theorem.

3.4.3 Theorem. [Dimension for a Tower Theorem] Consider a tower of subfields of $\mathbb{C}$,

$$\mathbb{E} \subseteq \mathbb{F} \subseteq \mathbb{K}.$$

If the vector spaces $\mathbb{F}$ over $\mathbb{E}$ and $\mathbb{K}$ over $\mathbb{F}$ have finite dimension then so does $\mathbb{K}$ over $\mathbb{E}$ and

$$[\mathbb{K} : \mathbb{E}] = [\mathbb{K} : \mathbb{F}][\mathbb{F} : \mathbb{E}]$$

or, in alternative notation,

$$\dim_{\mathbb{E}} \mathbb{K} = \dim_{\mathbb{F}} \mathbb{K} \cdot \dim_{\mathbb{E}} \mathbb{F}.$$

Proof. There are altogether mn of the $\alpha\beta$'s in Theorem 3.4.1 and these form a basis for $\mathbb{K}$ over $\mathbb{E}$. This vector space therefore has dimension mn. ■

Since the dimension of a vector space must be an integer, Theorem 3.4.3 has the following corollary.

3.4.4 Corollary. *Under the conditions of Theorem 3.4.3,* $[\mathbb{K} : \mathbb{F}]$ *is a factor of* $[\mathbb{K} : \mathbb{E}]$ *and* $[\mathbb{F} : \mathbb{E}]$ *is a factor of* $[\mathbb{K} : \mathbb{E}]$. ■

As an application of this corollary, observe that there cannot be any field $\mathbb{F}$ such that

$$\mathbb{Q} \subseteq \mathbb{F} \subseteq \mathbb{Q}(\sqrt{2})(\sqrt{3})$$

with $[\mathbb{F} : \mathbb{Q}] = 3$ since, by Example 3.4.2,

$$[\mathbb{Q}(\sqrt{2})(\sqrt{3}) : \mathbb{Q}] = 4$$

and 3 is not a factor of 4.

Exercises 3.4

1. How many vector spaces can be constructed from a tower of four distinct fields

$$\mathbb{E} \subseteq \mathbb{F} \subseteq \mathbb{G} \subseteq \mathbb{K}$$

by using larger fields as vector spaces over smaller fields?

2. Consider a tower of subfields of $\mathbb{C}$,

$$\mathbb{E} \subseteq \mathbb{F} \subseteq \mathbb{K}.$$

   (a) What is $[\mathbb{K} : \mathbb{E}]$ if $[\mathbb{K} : \mathbb{F}] = 2$ and $[\mathbb{F} : \mathbb{E}] = 6$?
   (b) Is it possible that $[\mathbb{K} : \mathbb{E}] = 3$ and $[\mathbb{F} : \mathbb{E}] = 2$? (Justify your answer.)

3. Give an example of a tower of three distinct subfields of $\mathbb{C}$,

$$\mathbb{E} \subseteq \mathbb{F} \subseteq \mathbb{K},$$

   such that $[\mathbb{K} : \mathbb{E}] = 4$.

4. By using the tower

$$\mathbb{Q} \subseteq \mathbb{Q}(\sqrt{3}) \subseteq \mathbb{Q}(\sqrt{3})(\sqrt{2})$$

   write down a basis for the vector space

$$\mathbb{Q}(\sqrt{3})(\sqrt{2}) \text{ over } \mathbb{Q}.$$

Exercises 5 to 10 will refer to the tower

$$\mathbb{Q} \subseteq \mathbb{Q}(\sqrt{6}) \subseteq \mathbb{Q}(\sqrt{2}+\sqrt{3}).$$

5. (a) Is the set $\{1, \sqrt{2}, \sqrt{3}, \sqrt{6}\}$ linearly independent over $\mathbb{Q}$? Why? [Hint. Use your answer to Exercise 4.]
   (b) Does $\sqrt{2}+\sqrt{3} \in \mathbb{Q}(\sqrt{6})$? Why?
   (c) Does $\sqrt{6} \in \mathbb{Q}(\sqrt{2}+\sqrt{3})$? Why? [Hint. Calculate $(\sqrt{2}+\sqrt{3})^2$.]
6. Which of your answers to Exercise 5 would you use
   (i) to validate the above tower?
   (ii) to show that $\mathbb{Q}(\sqrt{6})(\sqrt{2}+\sqrt{3}) = \mathbb{Q}(\sqrt{2}+\sqrt{3})$?
7. (a) Show that the number $\sqrt{2}+\sqrt{3}$ is algebraic over $\mathbb{Q}(\sqrt{6})$.
   (b) Write down $\mathrm{irr}(\sqrt{2}+\sqrt{3}, \mathbb{Q}(\sqrt{6}))$ and state which fact from Exercise 5 you would need to use to justify your answer.
   (c) Deduce the value of $[\mathbb{Q}(\sqrt{6})(\sqrt{2}+\sqrt{3}) : \mathbb{Q}(\sqrt{6})]$.
   (d) Deduce the value of $[\mathbb{Q}(\sqrt{2}+\sqrt{3}) : \mathbb{Q}(\sqrt{6})]$.
8. Use your answer to Exercise 7(d) and the Dimension for a Tower Theorem 3.4.3 to deduce
   (i) $[\mathbb{Q}(\sqrt{2}+\sqrt{3}) : \mathbb{Q}]$,
   (ii) a basis for the vector space $\mathbb{Q}(\sqrt{2}+\sqrt{3})$ over $\mathbb{Q}$.
9. Let $\beta \in \mathbb{Q}(\sqrt{2}+\sqrt{3})$.

(a) Use the fact that any $m+1$ vectors in a vector space of dimension m are linearly dependent, and your answer to Exercise 8(i), to show that $\{1, \beta, \beta^2, \beta^3, \beta^4\}$ is linearly dependent over $\mathbb{Q}$.
(b) Deduce from (a) that β is algebraic over $\mathbb{Q}$.
(c) What can you say about $\deg(\beta, \mathbb{Q})$?

10. (a) Use your answer to Exercise 8(ii) and the dimension part of the Basis for $F(\alpha)$ Theorem 3.2.4 to deduce the value of $\deg(\sqrt{2}+\sqrt{3}, \mathbb{Q})$.
[Hint. Remember that any two bases of a vector space contain the same number of vectors.]
(b) Use (a) and the Basis for $\mathbb{F}(\alpha)$ Theorem 3.2.4 to write down a basis for $\mathbb{Q}(\sqrt{2}+\sqrt{3})$ over $\mathbb{Q}$. Is the fact that this is different from the one in Exercise 8(ii) a problem?

Exercises 11 to 15 will refer to the tower

$$\mathbb{Q} \subseteq \mathbb{Q}(\sqrt{2}) \subseteq \mathbb{Q}(\sqrt[4]{2}).$$

11. (a) Show that $\sqrt[4]{2}$ is algebraic over $\mathbb{Q}$. Is it also algebraic over $\mathbb{Q}(\sqrt{2})$?
(b) Does $\sqrt[4]{2} \in \mathbb{Q}(\sqrt{2})$? (Justify your answer.)
(c) Does $\sqrt{2} \in \mathbb{Q}(\sqrt[4]{2})$? (Justify your answer.)

12. Which of your answers to Exercise 11 would you use
(ii) to validate the above tower?
(ii) to show that $\mathbb{Q}(\sqrt{2})(\sqrt[4]{2}) = \mathbb{Q}(\sqrt[4]{2})$?

13. (a) Write down the polynomial $\mathrm{irr}(\sqrt[4]{2}, \mathbb{Q}(\sqrt{2}))$ and state which fact from Exercise 11 you would use to justify your answer.
(b) Deduce the value of $[\mathbb{Q}(\sqrt{2})(\sqrt[4]{2}) : \mathbb{Q}(\sqrt{2})]$.
(c) Deduce the value of $[\mathbb{Q}(\sqrt[4]{2}) : \mathbb{Q}(\sqrt{2})]$.

14. Use your answer to Exercise 13(c) and the Dimension for a Tower Theorem 3.4.3 to deduce
(i) the value of $[\mathbb{Q}(\sqrt[4]{2}) : \mathbb{Q}]$,
(ii) a basis for the vector space $\mathbb{Q}(\sqrt[4]{2})$ over $\mathbb{Q}$,
(iii) the polynomial $\mathrm{irr}(\sqrt[4]{2}, \mathbb{Q})$.

15. Let $\mathbb{F}$ be a subfield of $\mathbb{C}$ and let α be a real number which is algebraic over $\mathbb{F}$ of degree n.

(a) What is $[\mathbb{F}(\alpha) : \mathbb{F}]$?
(b) What is the maximum number of vectors in a linearly independent subset of the vector space $\mathbb{F}(\alpha)$ over $\mathbb{F}$?
(c) Assume now that $\beta \in \mathbb{F}(\alpha)$.
 (i) Show that β is algebraic over $\mathbb{F}$.

(ii) Use a suitable tower to prove that the degree of β over $\mathbb{F}$ is a factor of that of α over $\mathbb{F}$.

(d) Use your answer to (c) to show that every number in $\mathbb{Q}(\sqrt[4]{2})$ is algebraic over $\mathbb{Q}$ and has degree 1, 2 or 4 over $\mathbb{Q}$.

Additional Reading for Chapter 3

Our treatment of $\mathbb{F}(\alpha)$ uses relatively little algebraic machinery. Other treatments rely on the algorithm for finding the greatest common divisor of two polynomials (see, for example, (Clark, 1971, §110), (Hadlock, 1978, §2.2), or (Shapiro, 1975, §10.1)) or they rely on results about maximal ideals in polynomial rings (see, for example, (Fraleigh, 1982, §35), or (Gilbert, 1976, Chapter 11)). The proof of the Basis for a Tower Theorem 3.4.1 is quite standard and can also be found in (Clark, 1971, §97), (Fraleigh, 1982, §38), and (Hadlock, 1978, §2.2).

Chapter 4
Irreducible Polynomials

In our earlier definition of the irreducible polynomial of a number, the word "irreducible" was intended to convey the idea that the degree of the polynomial "could not be reduced further". In this chapter it will be shown that this polynomial is also "irreducible" in the sense that it "cannot be factorized further". This will lead to a practical technique for finding the irreducible polynomial of a number.

With the earlier definition we were able to show, for example, that the irreducible polynomial of $\sqrt{2}$ over $\mathbb{Q}$ is $X^2 - 2$. The techniques to be developed in this chapter will enable us to move beyond quadratics and to show, for example, that

$$\mathrm{irr}(\sqrt[3]{2}, \mathbb{Q}) = X^3 - 2.$$

This will feature later in our proof of the impossibility of doubling the cube.

4.1 Irreducible Polynomials

In this section the concept of an "irreducible polynomial" will be defined. We have already met this phrase in the combination "the irreducible polynomial of a number α". Whereas the earlier concept was defined in terms of "least degree", we shall now turn to the idea of "cannot be further factorized". The precise relationship between an

"irreducible polynomial"

and

"the irreducible polynomial of a number α"

will be explained later, in Section 4.3. For the moment, however, the main thing is that you do not confuse the two concepts.

S. A. Morris et al., *Abstract Algebra and Famous Impossibilities*, Undergraduate Texts in Mathematics, https://doi.org/10.1007/978-3-031-05698-7_4

4.1.1 Definition. Let $\mathbb{F}$ be a field. A polynomial $f(X) \in \mathbb{F}[X]$ is said to be *reducible over* $\mathbb{F}$ if there are polynomials $g(X)$ and $h(X)$ in $\mathbb{F}[X]$ such that

(i) each has degree less than that of $f(X)$, and
(ii) $f(X) = g(X)h(X)$. ■

Thus a polynomial is reducible if it can be factorized in a suitable way.

4.1.2 Example. The polynomial $X^2 - 2$ is reducible over $\mathbb{R}$.

Proof. Observe that $X^2 - 2 = (X - \sqrt{2})(X + \sqrt{2})$, where each of the factors belongs to $\mathbb{R}[X]$ and has degree less than that of $X^2 - 2$. ■

4.1.3 Example. The polynomial $X^2 - 2$ is not reducible over $\mathbb{Q}$.

Proof. To see this, suppose, on the contrary, that $X^2 - 2$ is reducible over $\mathbb{Q}$. This means that

$$X^2 - 2 = (aX + b)(cX + d)$$

for some $a, b, c, d \in \mathbb{Q}$. Clearly neither a nor c can be zero. Evaluating both sides at $\sqrt{2}$ gives

$$0 = (a\sqrt{2} + b)(c\sqrt{2} + d).$$

Hence $\sqrt{2} = -b/a$ or $-d/c$, which gives the contradiction that $\sqrt{2}$ is rational. ■

4.1.4 Example. Constant polynomials and polynomials of degree 1 are not reducible. ■

Linguistically speaking, the word "irreducible" should simply mean not reducible, because the prefix ir means not.
(For example,
irreplaceable means "not replaceable",
irresponsible means "not responsible",
irreligious means "not religious".)
This is nearly true in the following definition, but there is a slight twist in that constant polynomials are excluded.

4.1.5 Definition. Let $\mathbb{F}$ be a field. A polynomial $f(X)$ in $\mathbb{F}[X]$ is said to be *irreducible over* $\mathbb{F}$ if it is not reducible over $\mathbb{F}$ and it is not a constant. ■

Thus an irreducible polynomial is a non-constant polynomial that cannot be factored non-trivially. The reason for excluding constant polynomials in this definition is that, without this exclusion, certain theorems which appear later would need to be stated in a more cumbersome way.

In terms of "irreducible", the examples given earlier show that

$$X^2 - 2 \quad \textit{is not irreducible over } \mathbb{R},$$
$$X^2 - 2 \quad \textit{is irreducible over } \mathbb{Q}.$$

Note that if a polynomial is irreducible over a field, then it is irreducible over all subfields of that field. More precisely, if $\mathbb{E}$ is a subfield of $\mathbb{F}$ and $f(X) \in \mathbb{E}[X]$ is irreducible over $\mathbb{F}$, then $f(X)$ is irreducible over $\mathbb{E}$.

Notice also that constant polynomials are neither reducible nor irreducible; but all other polynomials are either reducible or irreducible over a given field.

Exercises 4.1

1. In each case state (giving reasons) whether the polynomial is reducible, irreducible or neither over the given field.
   (a) the polynomial $X^2 + 3X + 2$ (i) over $\mathbb{R}$, (ii) over $\mathbb{Q}$;
   (b) the polynomial $X^2 + 1$
   (i) over $\mathbb{Q}$, (ii) over $\mathbb{C}$, (iii) over $\mathbb{Z}_2$, (iv) over $\mathbb{Z}_3$;
   (c) the polynomial 10 over $\mathbb{Q}$;
   (d) the polynomial $X^2 - 2$ (i) over $\mathbb{Q}$, (ii) over $\mathbb{Q}(\sqrt{2})$.

2. Does the fact that $3X^2 + 3 = 3(X^2 + 1)$ in $\mathbb{Q}[X]$ mean that $3X^2 + 3$ is reducible over $\mathbb{Q}$?

3. (a) Is $X^3 - 1$ irreducible over $\mathbb{Q}$?
   (b) Is $X^3 + 1$ irreducible over $\mathbb{Q}$?
   (c) Is $X^{666} - 1$ irreducible over $\mathbb{Q}$?

4. Let $f(X) \in \mathbb{F}[X]$ be irreducible over $\mathbb{F}$ and let $g(X), h(X)$ in $\mathbb{F}[X]$ be such that $f(X) = g(X)h(X)$. Show that either $g(X)$ or $h(X)$ is a constant polynomial.

5. Assume that $\alpha \in \mathbb{C}$ is algebraic over a field $\mathbb{F} \subseteq \mathbb{C}$ and that $f(X) = \mathrm{irr}(\alpha, \mathbb{F})$. Show that $f(X)$ is not reducible over $\mathbb{F}$.
   [Hint. Use the fact that $f(X)$ is of least possible degree amongst the polynomials in $\mathbb{F}[X]$ with α as a zero, and the fact that if $pq = 0$ for p, q in $\mathbb{C}$ then either $p = 0$ or $q = 0$.]

4.2 Reducible Polynomials and Zeros

In this section the relationship between a polynomial having a zero and being reducible will be explored. This relationship will form the basis of our technique for proving the irreducibility of certain polynomials.

As a first step in this direction, the relationship between having a zero and having a factor of degree 1 will be explored.

4.2.1 Definition. Let $\mathbb{F}$ be a field. A polynomial $f(X) \in \mathbb{F}[X]$ is said to have a *factor of degree* 1 *in* $\mathbb{F}[X]$ if

$$f(X) = (aX + b)g(X)$$

where $a, b \in \mathbb{F}$ with $a \neq 0$ and where $g(X) \in \mathbb{F}[X]$. ■

4.2.2 Theorem. [Factor Theorem] *Let $\mathbb{F}$ be a field. A polynomial $f(X) \in \mathbb{F}[X]$ has a factor of degree 1 in $\mathbb{F}[X]$ if and only if $f(X)$ has a zero in $\mathbb{F}$.*

Proof. Assume firstly that $f(X)$ has a factor of degree 1. With notation as in Definition 4.2.1, it follows that

$$-b/a \in \mathbb{F}$$

is a zero of $f(X)$ since

$$\begin{aligned} f(-b/a) &= (a(-b/a) + b)\, g(-b/a) \\ &= 0. \end{aligned}$$

Conversely, assume that $\alpha \in \mathbb{F}$ is a zero of $f(X)$.

We aim to show that $X - \alpha$ is a factor of $f(X)$.

To do this we shall find a polynomial $q(X)$ such that $f(X) = (X - \alpha)q(X)$.

But how will we find such a $q(X)$?

We divide $f(X)$ by $X - \alpha$, which will give us a polynomial $q(X)$ plus a remainder, by the Division Theorem 1.3.2.

Our task then is to show that the remainder is 0.

If we divide $f(X)$ by $X - \alpha$ then (by the Division Theorem 1.3.2) there exist $q(X), r(X)$ in $\mathbb{F}[X]$ with

$$f(X) = (X - \alpha)q(X) + r(X) \tag{1}$$

and

$$\text{either } r(X) = 0 \text{ or } \deg r(X) < \deg(X - \alpha) = 1.$$

Since $r(X) = 0$ or its degree is less than 1, $r(X)$ must be a constant polynomial $c \in \mathbb{F} \subseteq \mathbb{F}[X]$, which means we can rewrite (1) as

$$f(X) = (X - \alpha)q(X) + c. \tag{2}$$

If we substitute $X = \alpha$ into this and use the fact that $f(\alpha) = 0$ (why?), we have

$$0 = f(\alpha) = (\alpha - \alpha)q(\alpha) + c = 0 + c = c.$$

Thus, from (2), $f(X) = (X - \alpha)q(X)$ which means $f(X)$ has the factor $X - \alpha$ of degree 1 in $\mathbb{F}[X]$. ■

The most useful theorem for dealing with polynomials of degree 2 or 3 is the following one.

4.2.3 Theorem. [Small Degree Irreducibility Theorem] *Let $\mathbb{F}$ be any field. Let $f(X)$ in $\mathbb{F}[X]$ have degree 2 or 3. Then $f(X)$ is reducible over $\mathbb{F}$ if and only if $f(X)$ has a zero in $\mathbb{F}$.*

Proof. Let $f(X) \in \mathbb{F}[X]$ have degree 2 or 3.

Assume firstly that $f(X)$ is reducible over $\mathbb{F}$, so that

$$f(X) = g(X)h(X)$$

for some nonconstant polynomials $g(X), h(X) \in \mathbb{F}[X]$. Since the degrees of $g(X)$ and $h(X)$ must add up to 2 or 3, one (or both) of these degrees must be 1. Hence, by the Factor Theorem 4.2.2, one of them must have a zero in $\mathbb{F}$; hence $f(X)$ must also have a zero in $\mathbb{F}$.

To prove the converse is easy: assume $f(X)$ has a zero in $\mathbb{F}$ and then apply Theorem 4.2.2. ■

4.2.4 Example. The polynomial $2X^3 - 5$ is irreducible over $\mathbb{Q}$.

Proof. By the Rational Roots Test 1.4.1 the only possible zeros in $\mathbb{Q}$ of the polynomial are

$$\pm 1, \ \pm \frac{1}{2}, \ \pm \frac{5}{2}, \ \pm 5,$$

none of which works (that is, none gives zero when substituted into $2X^3 - 5$).

Thus the polynomial has no zeros in $\mathbb{Q}$. Since its degree is 3, the Small Degree Irreducibility Theorem 4.2.3 is applicable and shows that $f(X)$ is irreducible over $\mathbb{Q}$. ■

It is important to observe that the Small Degree Irreducibility Theorem 4.2.3 would not be true if we removed the restriction about "degree 2 or 3". For example, the quartic

$$(X^2 + 1)(X^2 + 4)$$

is reducible over $\mathbb{Q}$ but has no zero in $\mathbb{Q}$.

Fortunately we do not need to worry about the irreducibility of quartics (or higher degree polynomials) in order to prove the impossibility of the three famous geometric constructions.

Exercises 4.2

1. Use the Rational Roots Test 1.4.1 and either the Factor Theorem 4.2.2 or the Small Degree Irreducibility Theorem 4.2.3 to decide which of the polynomials listed below is irreducible over the stated field:
   (a) $X^3 - 2X + 1$ over $\mathbb{Q}$,
   (b) $X^3 + 2X + 1$ over $\mathbb{Q}$,
   (c) $X^3 - 5$ over $\mathbb{Q}$.

2. In which of the following cases can the technique of Exercise 1 be used to decide if the polynomial is irreducible over the given field?
   (a) $X^4 + 2X^3 + 2X^2 + 2X + 1$ over $\mathbb{Q}$,
   (b) $X^4 + 2X^3 - 2X^2 + 2X + 1$ over $\mathbb{Q}$,
   (c) $X^3 - \sqrt{2}X + 1$ over $\mathbb{Q}$,
   (d) $X^2 - 2$ over $\mathbb{Q}$,
   (e) $X^2 - 2$ over $\mathbb{Q}(\sqrt{2})$.

3. In each case find a zero in $\mathbb{C}$ of the given polynomial and then use the method of proof of the Factor Theorem 4.2.2 to write down a factor of degree 1 in $\mathbb{C}[X]$ of the given polynomial.
   (a) $10X^2 + 10$,
   (b) $X^2 - 4X + 5$.

4. In each case give an example of a field $\mathbb{F}$ and a polynomial $f(X)$ in $\mathbb{F}[X]$ such that $f(X)$ has no zeros in $\mathbb{F}$ and yet $f(X)$ is not irreducible over $\mathbb{F}$.
   (a) $f(X)$ is constant.
   (b) $f(X)$ is not constant.

   Does this contradict the Small Degree Irreducibility Theorem 4.2.3?

5. In each case decide whether the statement is true or false, where $\mathbb{F}$ is a field. (Justify your answers.)
   (a) If $f(X) \in \mathbb{F}[X]$ is reducible over $\mathbb{F}$ then $f(X)$ has a zero in $\mathbb{F}$.
   (b) If $f(X) \in \mathbb{F}[X]$ has a zero in $\mathbb{F}$ then $f(X)$ is reducible over $\mathbb{F}$.
   (c) If $f(X) \in \mathbb{F}[X]$ has degree 4 and has a zero in $\mathbb{F}$, then $f(X)$ is reducible over $\mathbb{F}$.
   (d) If $f(X) \in \mathbb{F}[X]$ has degree 2 or more and has a zero in $\mathbb{F}$, then $f(X)$ is reducible over $\mathbb{F}$.

6. In each case decide whether the statement is true or false, where $\mathbb{F}$ is a field. (Justify your answers.)
   (a) If $f(X) \in \mathbb{F}[X]$ has degree 6 and is irreducible over $\mathbb{F}$, then $f(X)$ does not have a zero in $\mathbb{F}$.
   (b) If $f(X) \in \mathbb{F}[X]$ is irreducible over $\mathbb{F}$ then $f(X)$ does not have a zero in $\mathbb{F}$.

(c) If $f(X) \in \mathbb{F}[X]$ has degree 2 or more and is irreducible over $\mathbb{F}$, then $f(X)$ does not have a zero in $\mathbb{F}$.

4.3 Irreducibility and irr(α, $\mathbb{F}$)

The purpose of this section is to state and prove a theorem which links the idea of "irreducible polynomial" (as in Definition 4.1.5) to the idea of "irreducible polynomial of a number" (as in Definition 2.3.2).

4.3.1 Theorem. [Monic Irreducible Zero Theorem] *Let $\mathbb{F} \subseteq \mathbb{C}$ be a field and let $\alpha \in \mathbb{C}$ be algebraic over $\mathbb{F}$. The following conditions on a polynomial $f(X) \in \mathbb{F}[X]$ are equivalent:*

(i) $f(X) = \text{irr}(\alpha, \mathbb{F})$,
(ii) $f(\alpha) = 0$ *and* $f(X)$ *is monic and irreducible over* $\mathbb{F}$.

Before proving this theorem, we give some examples to show how the theorem can be used to check that a guess for irr(α, $\mathbb{F}$) is indeed the correct one.

4.3.2 Example. $\text{irr}(\sqrt{2}, \mathbb{Q}) = X^2 - 2$.

Proof. Clearly $X^2 - 2$ has $\sqrt{2}$ as a zero and it is monic.

By the Rational Roots Test 1.4.1, the only possible zeros in $\mathbb{Q}$ are $\pm 1, \pm 2$, none of which works. Because this polynomial has degree 2, it follows from the Small Degree Irreducibility Theorem that it is irreducible over $\mathbb{Q}$. The Monic Irreducible Zero Theorem 4.3.1 now gives the required result. ■

Actually, this example was done earlier (as Example 2.3.3). The advantage of the Monic Irreducible Zero Theorem 4.3.1 is that it lets us handle cubics as well.

4.3.3 Example. $\text{irr}(\sqrt[3]{2}, \mathbb{Q}) = X^3 - 2$.

Proof. Clearly $X^3 - 2$ has $\sqrt[3]{2}$ as a zero and it is monic.

By the Rational Roots Test 1.4.1, the only possible zeros in $\mathbb{Q}$ are $\pm 1, \pm 2$, none of which works. Because this polynomial has degree 3, it follows from the Small Degree Irreducibility Theorem 4.2.3 that it is irreducible over $\mathbb{Q}$. The Monic Irreducible Zero Theorem 4.3.1 now gives the desired result. ■

Note that the same argument could not be used for quartics because of the requirement on the degree of the polynomial contained in the Small Degree Irreducibility Theorem 4.2.3. Note also that since we are appealing to the Rational Roots Test 1.4.1, the above technique for finding irr(α, $\mathbb{F}$) works only in the case $\mathbb{F} = \mathbb{Q}$.

We now complete this section by proving the theorem.

Proof of Theorem 4.3.1. To facilitate the proof, the following notation is introduced:

$$\mathcal{P} = \{p(X) \in \mathbb{F}[X] : p(\alpha) = 0 \text{ and } p(X) \text{ is monic}\}.$$

We firstly show that (i) implies (ii).

Assume that (i) is true. By Definition 2.3.2 this means that

$$f(X) \text{ is the polynomial of least degree in } \mathcal{P}.$$

Thus the first two statements in (ii) of the theorem hold and it remains to prove that $f(X)$ is irreducible over $\mathbb{F}$.

Suppose $f(X)$ were not irreducible over $\mathbb{F}$. Since it is not constant, it must be reducible over $\mathbb{F}$ and so there are polynomials $g(X)$ and $h(X)$ in $\mathbb{F}[X]$ such that

$$f(X) = g(X)h(X)$$

where the degrees of $g(X)$ and $h(X)$ are both less than that of $f(X)$. We may assume that both of these polynomials are monic. Evaluation of both sides at α gives, furthermore,

$$0 = g(\alpha)h(\alpha)$$

and so, $\mathbb{C}$ being a field, this implies $g(\alpha) = 0$ or $h(\alpha) = 0$. Thus either

$$g(X) \in \mathcal{P} \quad \text{or} \quad h(X) \in \mathcal{P}.$$

But this is a contradiction since $f(X)$ had the least degree in $\mathcal{P}$. Hence $f(X)$ is irreducible over $\mathbb{F}$. So (i) does imply (ii).

Next we show that (ii) implies (i).

Assume that (ii) is true. Thus $f(X) \in \mathcal{P}$ and so

$$\text{degree } f(X) \geq \text{ degree irr}(\alpha, \mathbb{F}).$$

By the Division Theorem 1.3.2,

$$f(X) = \text{irr}(\alpha, \mathbb{F})q(X) + r(X), \tag{1}$$

where $q(X)$ and $r(X) \in \mathbb{F}(X)$ with $r(X) = 0$ or

$$\text{degree } r(X) < \text{ degree irr}(\alpha, \mathbb{F}).$$

Evaluation of the equation (1) at α gives

$$0 = 0.q(\alpha) + r(\alpha)$$

so that $r(\alpha) = 0$. Suppose $r(X) \neq 0$. Then we can divide $r(X)$ by its leading coefficient to get a monic polynomial in $\mathcal{P}$ whose degree is less than that of $\text{irr}(\alpha, \mathbb{F})$, which is a contradiction. Hence $r(X) = 0$ and so by the equation (1),

$$f(X) = \text{irr}(\alpha, \mathbb{F})q(X).$$

Thus $q(X)$ must be a constant, since otherwise $f(X)$ would not be irreducible over $\mathbb{F}$. Since both $f(X)$ and $\operatorname{irr}(\alpha, \mathbb{F})$ are monic, this means $q(X)$ is 1. ■

Exercises 4.3

1. Find $\operatorname{irr}(\sqrt{5}+1, \mathbb{Q})$ and $\operatorname{irr}(\sqrt[3]{3}-2, \mathbb{Q})$ and then use the method of Examples 4.3.2 and 4.3.3 to justify your answer.
2. (a) Guess $\operatorname{irr}(\sqrt[3]{7}, \mathbb{Q})$ and $\deg(\sqrt[3]{7}, \mathbb{Q})$.
   (b) Now use the definition of $\mathbb{Q}(\sqrt[3]{7})$ to get an explicit description of this set.
   (c) Give a basis for the vector space $\mathbb{Q}(\sqrt[3]{7})$ over $\mathbb{Q}$.
   (d) Give a careful proof, with full justification for each step, for your answer to part (a).
3. Suppose that $\sqrt{7} \in \mathbb{Q}(\sqrt[3]{7})$.
   (a) Deduce that $\mathbb{Q}(\sqrt{7}) \subseteq \mathbb{Q}(\sqrt[3]{7})$.
   (b) Thus we have a tower $\mathbb{Q} \subseteq \mathbb{Q}(\sqrt{7}) \subseteq \mathbb{Q}(\sqrt[3]{7})$.
   Derive a contradiction from this tower.
   (c) What can you now conclude?
4. Note the obvious tower

$$\mathbb{Q} \subseteq \mathbb{Q}(\sqrt[3]{7}) \subseteq \mathbb{Q}(\sqrt[3]{7})(\sqrt{7}).$$

   (a) Guess $\operatorname{irr}(\sqrt{7}, \mathbb{Q}(\sqrt[3]{7}))$ and $\deg(\sqrt{7}, \mathbb{Q}(\sqrt[3]{7}))$.
   (b) Prove your answer to part (a) using the result of Exercise 3.
   (c) Hence give a basis for the two vector spaces

$$\mathbb{Q}(\sqrt[3]{7})(\sqrt{7}) \quad \text{over} \quad \mathbb{Q}(\sqrt[3]{7}),$$
$$\mathbb{Q}(\sqrt[3]{7})(\sqrt{7}) \quad \text{over} \quad \mathbb{Q},$$

   and state their dimensions.
5. In each of the following cases show that the number $\sin\theta$ is algebraic over $\mathbb{Q}$ and find its irreducible polynomial and its degree over this field.
   (a) $\theta = \pi/6$,
   (b) $\theta = \pi/3$,
   (c) $\theta = \pi/18$. [Hint. Use the identity $\sin(3\theta) = 3\sin\theta - 4\sin^3\theta$.]
6. Assume that $\mathbb{E}$ is an extension field of $\mathbb{Q}$, and $\alpha \in \mathbb{E}$ is such that

$$2\alpha^3 + 3\alpha^2 + 4\alpha + 3 = 0.$$

   (a) Write down a monic polynomial $f(X)$ in $\mathbb{Q}[X]$ such that α is a zero of $f(X)$.
   (b) Can we find $\operatorname{irr}(\alpha, \mathbb{Q})$ from the information given? Why?

4.4 Finite-dimensional Extensions

Let $\mathbb{K}$ be an extension field of $\mathbb{F}$ such that the vector space $\mathbb{K}$ over $\mathbb{F}$ is finite-dimensional. Must numbers in $\mathbb{K}$ be algebraic over $\mathbb{F}$ and, if so, what can we say about $\deg(\alpha, \mathbb{F})$ for α in $\mathbb{K}$? The following theorem answers these questions.

4.4.1 Theorem. *Let $\mathbb{F}$ be a subfield of a field $\mathbb{K}$ with $[\mathbb{K} : \mathbb{F}] = n$. Then every number α in $\mathbb{K}$ is algebraic over $\mathbb{F}$ and* $\deg(\alpha, \mathbb{F}) \leq n$.

Proof. Let $\alpha \in \mathbb{K}$. Because $[\mathbb{K} : \mathbb{F}] = n$, every set of $(n + 1)$ numbers in $\mathbb{K}$ must be linearly dependent over $\mathbb{F}$. Now

$$\{1, \alpha, \ldots, \alpha^n\}$$

is such a set and thus there exist $c_0, c_1, \ldots c_n \in \mathbb{F}$, not all zero, such that

$$c_0 1 + c_1\alpha + \ldots + c_n\alpha^n = 0.$$

If we let

$$p(X) = c_0 + c_1 X + \ldots + c_n X^n$$

then $p(X)$ is a nonzero polynomial in $\mathbb{F}[X]$ which has α as a zero. Thus (by definition), α is algebraic over $\mathbb{F}$. It follows easily (using Proposition 2.2.2) that $\deg(\alpha, \mathbb{F}) \leq n$ since $\deg p(X) \leq n$. ■

4.4.2 Corollary. *Let $\mathbb{F}$ be a subfield of $\mathbb{C}$ and let α be a complex number which is algebraic over $\mathbb{F}$ of degree n. Every number in $\mathbb{F}(\alpha)$ is then algebraic over $\mathbb{F}$ and has degree $\leq n$ over $\mathbb{F}$.* ■

4.4.3 Theorem. *Let $\mathbb{F}$ be a subfield of a field $\mathbb{E}$. The set of numbers in $\mathbb{E}$ which are algebraic over $\mathbb{F}$ is a subfield of $\mathbb{E}$.*

Proof. Let $\alpha, \beta \in \mathbb{E}$ be algebraic over $\mathbb{F}$. We must show that $\alpha + \beta$, $\alpha - \beta$, $\alpha\beta$ and (provided $\beta \neq 0$) α/β are all algebraic over $\mathbb{F}$. We consider the tower

$$\mathbb{F} \subseteq \mathbb{F}(\alpha) \subseteq \mathbb{F}(\alpha)(\beta).$$

Since α is algebraic over $\mathbb{F}$, $\mathbb{F}(\alpha)$ is a finite-dimensional extension of $\mathbb{F}$. Since β is algebraic over $\mathbb{F}$ it is also algebraic over $\mathbb{F}(\alpha)$ and hence $\mathbb{F}(\alpha)(\beta)$ is a finite-dimensional extension of $\mathbb{F}(\alpha)$. Thus, by the Dimension for a Tower Theorem 3.4.3, we see that $\mathbb{F}(\alpha)(\beta)$ is a finite-dimensional extension of $\mathbb{F}$ and so, by Theorem 4.4.1, every element of $\mathbb{F}(\alpha)(\beta)$ is algebraic over $\mathbb{F}$. This completes the proof since $\alpha + \beta$, $\alpha - \beta$, $\alpha\beta$ and (provided $\beta \neq 0$) α/β are all in $\mathbb{F}(\alpha)(\beta)$. ■

Recall that the algebraic numbers are those complex numbers which are algebraic over $\mathbb{Q}$.

4.4.4 Corollary. *The set $\mathbb{A}$ of all algebraic numbers is a subfield of $\mathbb{C}$.* ■

4.4.5 Definition. A field $\mathbb{F}$ is said to be *algebraically closed* if each polynomial $f(X) \in \mathbb{F}[X]$ has a zero in $\mathbb{F}$.

You are familiar with the Fundamental Theorem of Algebra which says that polynomial $f(X) = a_n X^n + a_{n-1}X^{n-1} + \cdots + a_1 X + a_0$, where each a_i is a complex number, $a_n \neq 0$, and $n \geq 1$, has a zero; that is, there exists a complex number X_0 such that $f(X_0) = 0$. This can be restated as follows:

4.4.6 Theorem. [Restatement of the Fundamental Theorem of Algebra 8.2.1] *The field $\mathbb{C}$ of all complex numbers is algebraically closed.*

4.4.7 Examples. Clearly As the polynomial $f(X) = X^2 - 1$ has no zero in $\mathbb{R}$, the field $\mathbb{R}$ and all its subfields (e.g. $\mathbb{Q}$) are not algebraically closed.

4.4.8 Theorem. *The field $\mathbb{A}$ of all algebraic numbers is algebraically closed.*

Proof. Let u be a zero of the polynomial $f(X) \in \mathbb{A}[X]$, where $f(X) = X^n + a_{n-1}X^{n-1} + a_{n-2}X^{n-2} + \cdots + a_1 X + a_0$ so that

$$u^n + a_{n-1}u^{n-1} + \cdots + a_0 = 0, \tag{1}$$

where $a_0, a_1, \ldots, a_{n-1} \in \mathbb{A}$.

Consider $\mathbb{K} = \mathbb{Q}(a_0, a_1, \ldots, a_{n-1})$, which is a finite-dimensional extension field of $\mathbb{Q}$. As u is a zero of $f(X)$, u is algebraic over $\mathbb{K}$. Therefore $\mathbb{K}(u)$ is a finite-dimensional extension field of $\mathbb{K}$ and hence, by Theorem 3.4.3, also of $\mathbb{Q}$. By Theorem 4.4.1 every element of $\mathbb{K}(u)$ is algebraic over $\mathbb{Q}$. In particular u is an algebraic number; that is, $u \in \mathbb{A}$. So we see by Definition 4.4.5, $\mathbb{A}$ is an algebraically closed field. ■

Exercises 4.4

1. Use Theorem 4.4.1 to show that every complex number z is algebraic over $\mathbb{R}$. What can you say about $\deg(z, \mathbb{R})$?
2. (a) Show that, with the hypotheses of Theorem 4.4.1, $\deg(\alpha, \mathbb{F})$ is a factor of n. (Consider the tower $\mathbb{F} \subseteq \mathbb{F}(\alpha) \subseteq \mathbb{K}$.)
   (b) If $\alpha \in \mathbb{Q}(\sqrt[4]{2})$, what can you say about $\deg(\alpha, \mathbb{Q})$?
3. Explain why the number $17\sqrt[4]{2} - 3\sqrt{2}$ is in the field $\mathbb{Q}(\sqrt[4]{2})$ and then use Theorem 4.4.1 to prove that this number is algebraic over the field $\mathbb{Q}$.
4. (a) Explain why $\mathbb{Q}(i)(\sqrt{2})$ is a finite-dimensional extension of $\mathbb{Q}$.
   (b) Use Theorem 4.4.1 to prove that $(3i + 2\sqrt{2})/(1 - 4\sqrt{2}i)$ is algebraic over $\mathbb{Q}$.

5.* Let $\mathbb{K}$ be an extension field of $\mathbb{R}$ such that every element α in $\mathbb{K}$ is algebraic and of degree at most 2 over $\mathbb{R}$. Prove that $\mathbb{K} = \mathbb{R}$ or $\mathbb{C}$.
[Hint. Assume $\deg(\alpha, \mathbb{R}) = 2$. By completing the square in $\mathrm{irr}(\alpha, \mathbb{R})$, show that there are real numbers b and c such that $\beta = (\alpha - b)/c$ satisfies $\beta^2 = -1$.]

6.* Suppose that there exists an extension field $\mathbb{K}$ of $\mathbb{R}$ such that $[\mathbb{K} : \mathbb{R}] = 3$.
(a) Prove that there exists an element α in $\mathbb{K}$ which is algebraic and of degree 3 over $\mathbb{R}$.
[Hint. Use Exercise 5.]
(b) Now note that every cubic polynomial in $\mathbb{R}[X]$ has a zero in $\mathbb{R}$. (Proof via the Intermediate Value Theorem, as in (Spivak, 1967, p. 103).
(c) Explain why (a) and (b) contradict the original supposition.

7. Verify that if $\mathbb{F}$ is an algebraically closed subfield of $\mathbb{C}$, then $\mathbb{F}$ contains all n^{th} roots of unity, for $n \in \mathbb{N}$.

[Remarks: *elementary vector algebra*. The set of all vectors in the plane can be regarded as the set $\mathbb{R}^2$ of all ordered pairs of real numbers (a, b). The vector space $\mathbb{R}^2$ can be made into a field by adopting the definition of multiplication from $\mathbb{C}$; thus

$$(a, b)(c, d) = (ac - bd, ad + bc).$$

Elements of the form $(a, 0)$ then form a "copy" of $\mathbb{R}$ in $\mathbb{R}^2$.

The set of vectors in three-dimensional space, however, behaves quite differently. This set can be regarded as the set $\mathbb{R}^3$ of all ordered triples of real numbers (a, b, c). Exercise 6 shows that it is impossible to make the vector space $\mathbb{R}^3$ into an extension field of $\mathbb{R}$, regardless of how we choose the vector multiplication in $\mathbb{R}^3$. Barehanded proofs of this result are included in Holden (1989) and Young (1988), while (Herstein, 1964, Chapter 7) contains a discussion of a more general result known as the Theorem of Frobenius.]

Additional Reading for Chapter 4

In Chapter 4 we developed techniques for proving that polynomials of degree 3 or less are irreducible over $\mathbb{Q}$. (Fraleigh, 1982, Example 31.6) will show you the sorts of computations that polynomials of degree 4 and higher can involve. There are also somewhat general techniques which are not restricted to such small degree polynomials and which apply over fields other than $\mathbb{Q}$. See (Fraleigh, 1982, §31.2), (Clark, 1971, §101–107), (Gilbert, 1976, Chapter 5), (Hadlock, 1978, §2.1), and (Shapiro, 1975, §2.6).

The problem of deciding, in general, whether a polynomial is irreducible over a given field is a difficult one. Rather deep techniques, amenable to computer use, have been developed. See Beardon and Ng (2000) and the chapter "Factorization of polynomials" by E. Kaltofen in Buchberger et al. (1983).

You may now wish to refer to other treatments of $\mathrm{irr}(\alpha, \mathbb{F})$ such as those given in (Fraleigh, 1982, §35), (Clark, 1971, §109) and (Hadlock, 1978, §2.2). Our Monic Irreducible Zero Theorem 4.3.1 is the link between our treatment and that in these books.

When discussing $\mathrm{irr}(\alpha, \mathbb{F})$, we have usually taken $\mathbb{F}$ to be a subfield of the complex number field $\mathbb{C}$ and α to be a complex number because this is all that our impossibility proofs in Chapter 5 require. You should note that nearly all of our results about $\mathrm{irr}(\alpha, \mathbb{F})$ generalize easily to the case where $\mathbb{F} \subseteq \mathbb{E}$ is any tower of fields and $\alpha \in \mathbb{E}$. Results about $\mathrm{irr}(\alpha, \mathbb{F})$ are stated in this more general form in Clark (1971) and Fraleigh (1982), for example.

Chapter 5
Straightedge and Compass Constructions

In the previous chapters we developed the algebraic machinery for proving that the three famous geometric constructions are impossible. In this chapter we introduce some geometry and start to show the connection between algebra and the geometry of constructions.

We begin by showing you how to do some basic constructions with straightedge and compass, and then how to construct line segments whose lengths are products, quotients, or square roots of ones already constructed.

Strict rules, passed down from the Greek geometers, must be followed in performing a construction. An important part of this chapter involves spelling out these rules very precisely. This leads to the notion of constructible number, on which the impossibility proofs in the next chapter are based.

We conclude the chapter by showing that there is a connection between geometric constructions and fields of the form $\mathbb{F}(\sqrt{\gamma})$.

5.1 Standard Straightedge and Compass Constructions

In this section we show you how to do some basic constructions with straightedge and compass. In each of these constructions, we are given some points and some lines passing through these points (and possibly some circles). We can construct new lines and circles using the straightedge and compass as described in 1 and 2 below.

1. The *straightedge* may be used to draw a new line, extended as far as we like, through any two points already in the figure.
2. The *compass* may be used to draw new circles in two ways.

   (a) Put the compass point on one point in the figure and the pencil on another such point and draw a circle or an arc of a circle. See Figure 5.1.

S. A. Morris et al., *Abstract Algebra and Famous Impossibilities*, Undergraduate Texts in Mathematics, https://doi.org/10.1007/978-3-031-05698-7_5

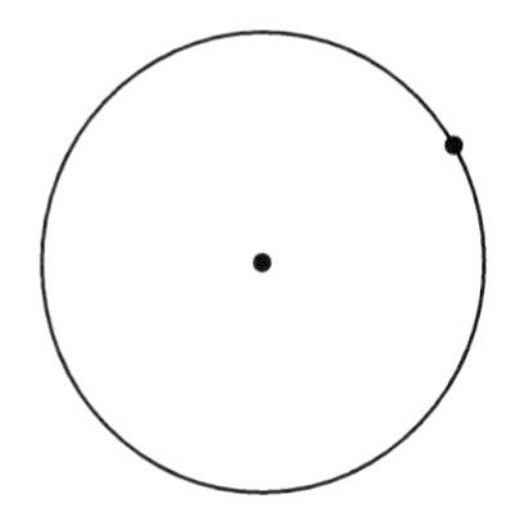

Fig. 5.1 Drawing a circle (**a**)

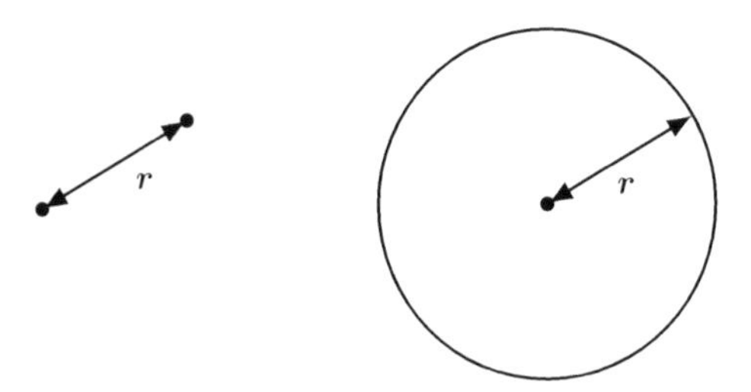

Fig. 5.2 Drawing a circle (**b**)

(b) Place the compass point and pencil as in (a) but then move the compass point to a third point in the figure before drawing the circle (or an arc) with this third point as centre. See Figure 5.2

> We get new points where the new line or circle meets other lines or circles.

In the rest of this section we show how you can use a straightedge and compass to perform several constructions. These constructions will be used later as building blocks for other constructions.

5.1.1 Bisecting a Line Segment

Purpose. *To construct the midpoint C of a given line segment AB.*

Method. See Figure 5.3.

(i) Put the compass point on A and extend the compass until its pencil is exactly on B. Now draw an arc above AB and another arc below AB.

(ii) Put the compass point on B and extend the compass until its pencil is exactly on A. Now draw arcs to intersect the arcs in (i). Call the points of intersection D and E respectively.
(iii) Join D and E with the straightedge. Where DE meets AB is the required point C.

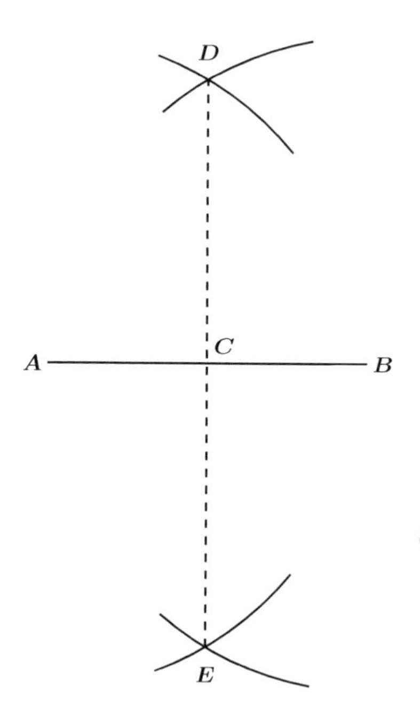

Fig. 5.3 Bisecting a line segment

[That C is the midpoint of AB seems obvious from the symmetry of the figure, and we give no formal proof of this (or of the validity of the constructions in the rest of this section). Readers familiar with Euclidean geometry proofs may be interested in providing their own proofs, using facts about congruent triangles.] ■

5.1.2 *Transferring a Length (Using a Compass)*

Purpose. *Given a line segment AB and a (longer) line segment CD, to construct a point P on CD such that the line segments AB and CP have equal lengths.*

Method. See Figure 5.4.

(i) Put the compass point on A and extend the compass until its pencil is exactly on B.
(ii) Put the compass point on C and with the same radius as in (i), draw an arc to cut CD at P. Then P is the required point.

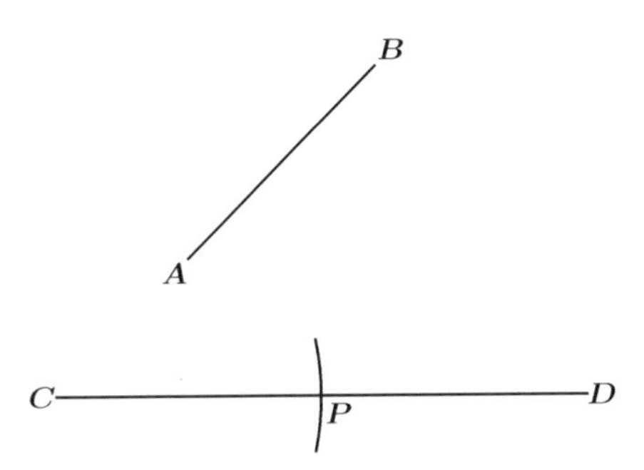

Fig. 5.4 Transferring a length

■

5.1.3 *Bisecting an Angle*

Purpose. *To construct a line OC which bisects a given angle* AOB.

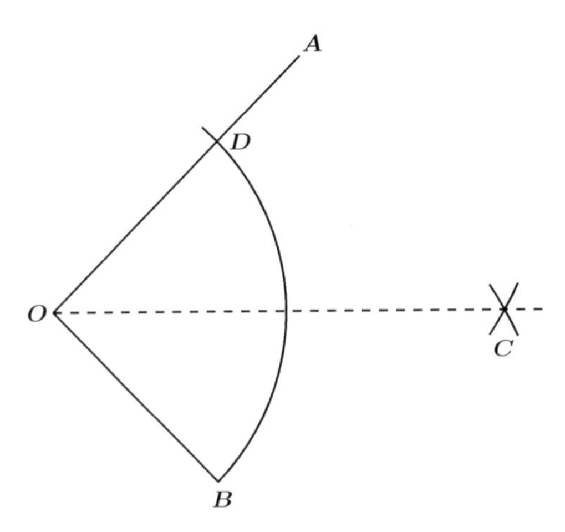

Fig. 5.5 Bisecting an angle

Method. See Figure 5.5.

(i) With centre O and radius OB, draw an arc to meet OA (extended if necessary) at D.
(ii) Draw arcs with centres D and B, and radius BD, to meet at a point C. Then OC bisects angle AOB. ■

5.1.4 *Constructing an Angle of* 60°

Purpose. *Given a line segment OA, to construct a line OB so that angle $AOB =$* 60°.

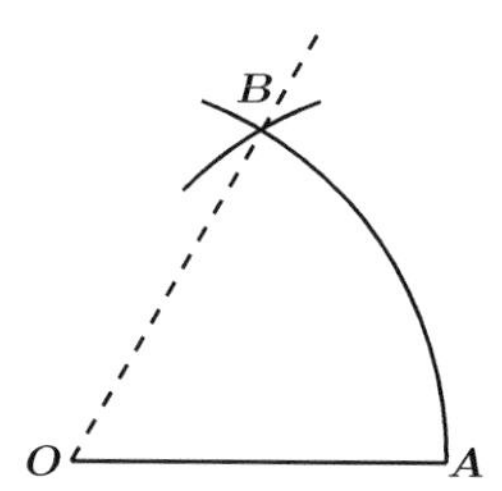

Fig. 5.6 Constructing an angle of 60°

Method. See Figure 5.6.
Draw arcs with centres O and A, and radius OA, which meet at a point B. Then angle $AOB = 60°$. ■

5.1.5 *Constructing an Angle of* 90°

Purpose. *Given a line segment OA, to construct a line OB so that angle $AOB =$* 90°.

Method. See Figure 5.7.

(i) With centre O draw an arc of radius OA which meets AO extended, at the point Q.
(ii) With centre A and radius AQ, draw arcs, one above the segment AQ and the other below AQ.

(iii) With centre Q and the same radius as in (ii), draw arcs to meet those in (ii) at points B and C respectively.
(iv) Join B and C, using the straightedge. This line passes through O and angle $AOB = 90°$.

[Note that steps (ii)–(iv) are the same as those for bisecting the line segment AQ.]

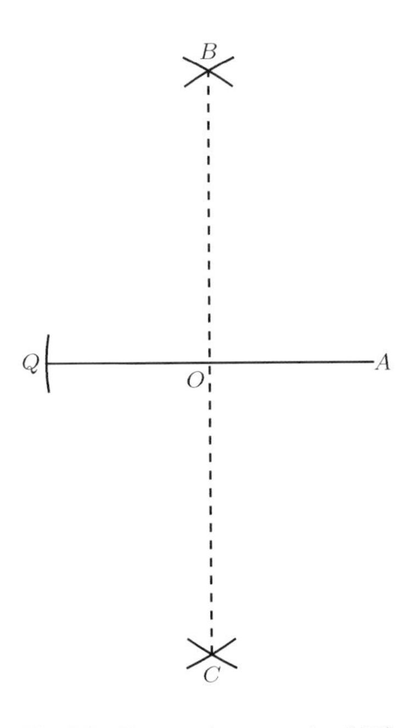

Fig. 5.7 Constructing an angle of 90°

■

5.1.6 Copying an Angle

Purpose. *Given an angle ABC and a line segment DE, to construct a line EF so that the angles ABC and DEF are equal.*

Method. See Figure 5.8.

(i) With centre B and radius BC draw an arc to meet AB (extended if necessary) at P.
(ii) With centre E and the same radius as in (i), draw an arc to meet ED (extended if necessary) at R.

(iii) With centre R and the radius PC, draw an arc to meet the arc drawn in (ii) at F. Then angles ABC and DEF are equal. ■

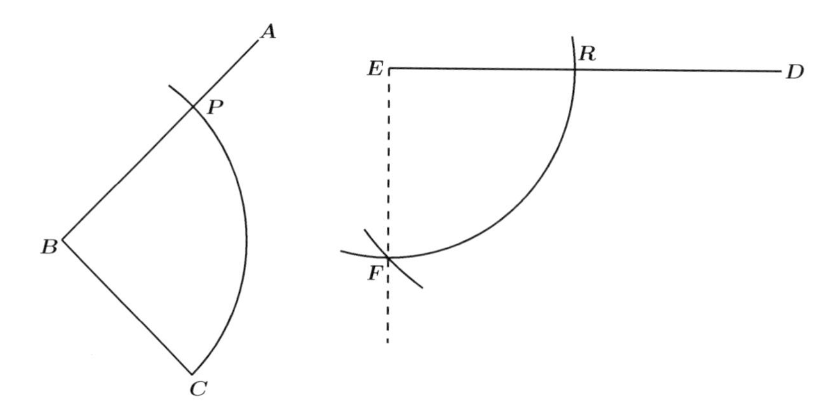

Fig. 5.8 Copying an angle

5.1.7 *Constructing a Line Parallel to a Given Line*

Purpose. *Given points A and B and a point C not on the line through A and B, to construct a line segment CD which is parallel to AB.*

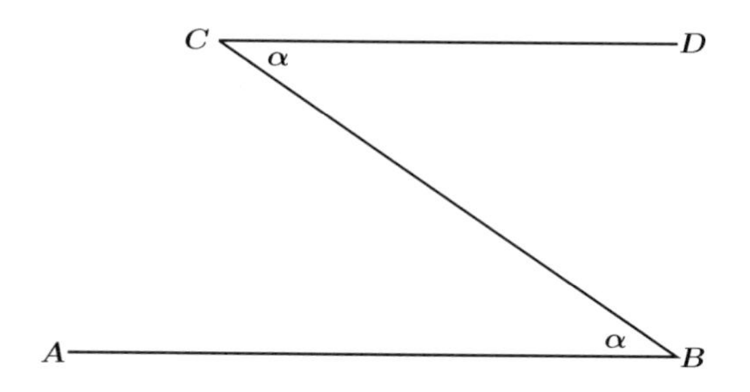

Fig. 5.9 Constructing a line parallel to a given line

Method. See Figure 5.9.

(i) Use the straightedge to draw the line joining C to B.

(ii) Use the angle copying construction given in 5.1.6 above to construct CD so that angles ABC and DCB are equal. Then CD is parallel to AB. ■

5.1.8 Use of Compass and Straightedge

In all of the above constructions, the straightedge has been used in strict conformity with the rules laid down by the ancient Greeks, who imagined the straightedge as being free of any markings. Thus, although it could be used to draw straight lines, it could not be used to measure distances between points or to transfer lengths.

Our use of the compass, on the other hand, calls for a more versatile compass than that allowed by the Greeks. The extra versatility lies in our assumption that the compass can be used to draw circles in both of the ways described in 2(a) and 2(b) at the start of Section 5.1, whereas the compass imagined by the Greeks could only be used in the first of these two ways. By allowing 2(b), we are assuming that our compass can be opened to a specified radius (as determined by the two points) and then picked up from the paper and used to draw a circle of this radius around a third point. We have used the compass this way in steps (i) and (ii) of 5.1.2 above, for example. The Greeks presumably thought of their compass as "collapsing" as soon as it was lifted from the paper and so it could not be used directly to transfer lengths.

It turns out, however, that it does not matter whether we use a *noncollapsing* compass (as in the constructions 5.1.2 and 5.1.6 above) or the apparently less powerful *collapsing* compass of the Greeks. It can be shown that

> any construction which can be performed with a noncollapsing compass can be performed also with a collapsing compass.

Use of the collapsing compass may, of course, involve taking more steps in the construction. For the details, see Exercise 10 below.

Exercises 5.1

1. Given two points A and B and a point C not on the line through A and B, show how to construct (using only straightedge and compass) a line through C which is perpendicular to AB. See Figure 5.10.

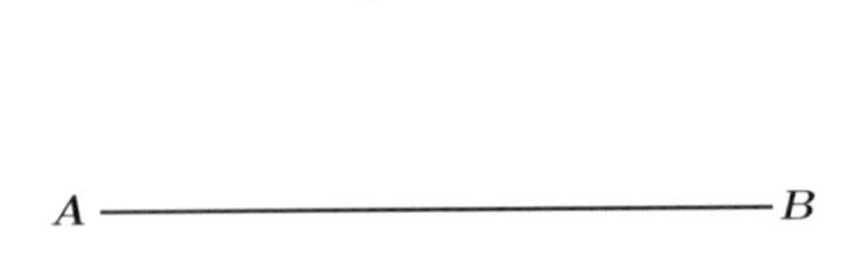

Fig. 5.10 Constructing a line through a point perpendicular to a line

2. (a) On a sheet of paper construct a line segment $A'B'$ having exactly the same length as AB shown below: See Figure 5.11.

Fig. 5.11 Constructing a line segment of length equal to a given line segment

   (b) Construct a line segment having one quarter of this length. For which positive integers n can you construct, in a similar way, a line segment whose length is $\frac{1}{n}$ times that of AB?

3. (a) On a sheet of paper construct an angle $A'O'B'$ which is equal to the angle AOB shown below: (See Figure 5.12.)

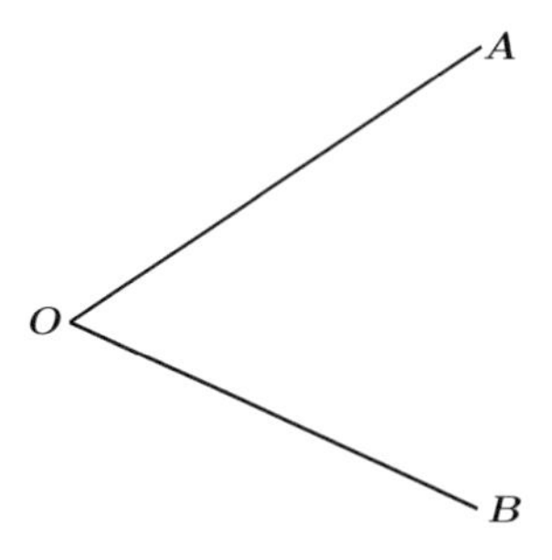

Fig. 5.12 Constructing an angle equal to a given angle

   (b) Construct an angle which is one quarter of this angle.
   (c) For which positive integers n can you construct, in a similar fashion, an angle which is $\frac{1}{n}$ times the angle AOB?

4. (a) Construct an equilateral triangle using straightedge and compass.
   (b) Which angle have you trisected in part (a)? How would you trisect an angle of $90°$?
   (c) Can you trisect, with straightedge and compass, an angle of $60°$? [Hint. Re-read Section 0.3.]

5. (a) Draw on a sheet of paper a line segment about 2 cm long and, taking this as your unit of length, construct line segments of lengths $\sqrt{3}$ and $1+\sqrt{3}$. [Hint. What is $\sin(60°)$?]
   (b) Appearances can be deceiving! Show that

$$\sqrt{4+2\sqrt{3}} = 1+\sqrt{3},$$
$$\sqrt{3+\sqrt{13+4\sqrt{3}}} = 1+\sqrt{3}.$$

[Thus you have already constructed in part (a) line segments having these horrible looking lengths.]

6. How would you construct a line segment of length $\sqrt{2}$?
7. Find positive integers a, b, c, d (different from those in Exercise 5(b)) such that $\sqrt{a+b\sqrt{3}} = c + d\sqrt{3}$.
8. (a) Given an angle AOB, describe how you would construct (using only straightedge and compass) an angle which is approximately equal to one third of angle AOB, with an error which is less than one eighth of the original angle.
   [Hint. Find an integer a with the fraction $a/8$ sufficiently close to 1/3.]
   (b) As for (a), but now require that the error is to be less than one sixteenth of the original angle.
   (c) Assume now that k is a positive integer. Describe a method of approximately trisecting an angle, using only straightedge and compass, which gives an error of at most 2^{-k} of the original angle.
9. Verify that the constructions 5.1.1, 5.1.3 and 5.1.4 can be performed exactly as described, using a collapsing compass.
10. First read through the construction set out below and then do (a) and (b) below.

    (a) Prove that the circle constructed at step (iii) has the required properties (thereby proving that the construction achieves its purpose).
    (b) Explain why the construction shows that *a collapsing compass can be used to draw any circle that a noncollapsing compass can draw.*

Purpose. *Given points O, A and B, to construct a circle with centre O and radius AB with a* collapsing *compass.*

Method. (See Figure 5.13.)

(i) Draw one circle with centre O and radius equal to the length of OA and another circle with centre A and radius OA. Let P and Q be the points where these two circles intersect.
(ii) Draw a circle with centre P and radius PB and another circle with centre Q and radius equal to QB. These circles intersect at B and also at another point which we call R.
(iii) Draw the circle with centre O and radius OR. (It is the dotted circle in our diagram.) This is the required circle.

11. Figure 5.14 below shows how to construct a line segment CD which is parallel to a line AB by using a "rusty compass", that is, one which cannot alter its radius. Why is CD parallel to AB?
    (For more on this fascinating topic see (Gardner, 1981, Chapter 17).)

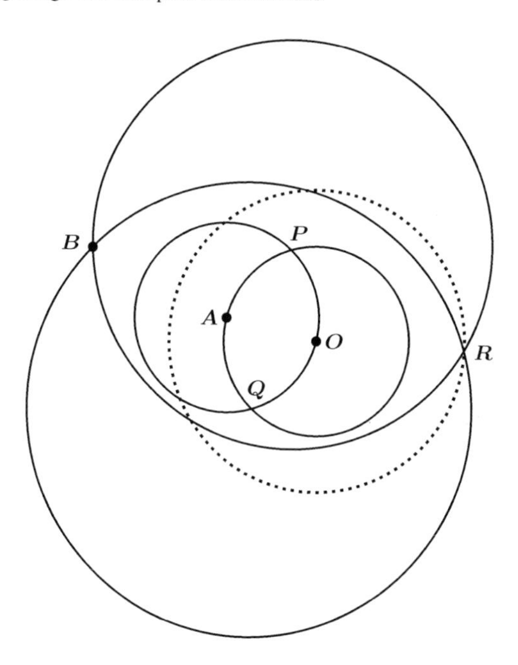

Fig. 5.13 Constructing a circle with a collapsing compass

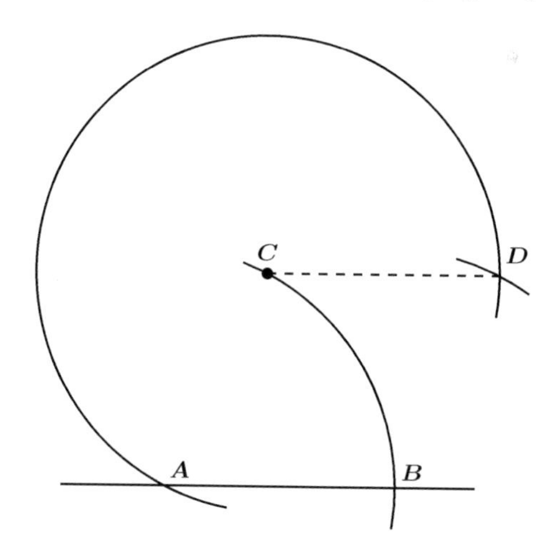

Fig. 5.14 Constructing a line segment with a rusty compass

5.2 Products, Quotients, Square Roots

5.2.1 Constructing a Product

Purpose. *Given line segments of lengths α and β, to construct a line segment of length $\alpha\beta$.*

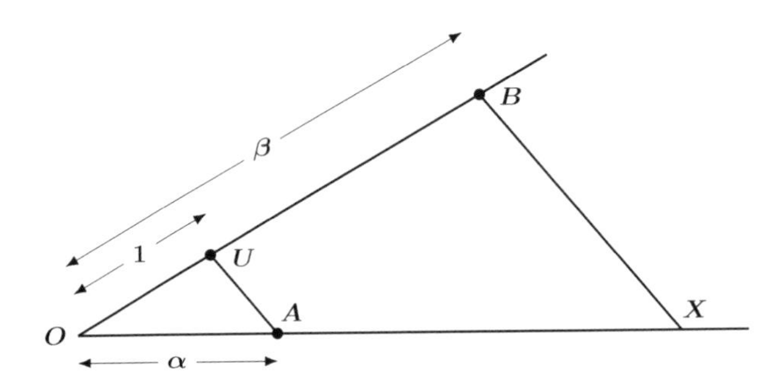

Fig. 5.15 Constructing a product

Method. (See Figure 5.15.)

(i) Draw two lines which intersect in a single point O.
(ii) On one of these lines, construct a point A such that the length of OA is α.
(iii) On the other line construct
  a point U such that the length of OU is 1, and
  a point B such that the length of OB is β.
(iv) Construct (as in 5.1.7 above) a line through B parallel to UA and meeting the line OA, extended if necessary, at X.

Result. *The line segment OX has length $\alpha\beta$.*

Proof. Let x be the length of OX. Because the triangles OAU and OXB are similar,

$$\frac{x}{\alpha} = \frac{\beta}{1}$$

and so

$$x = \alpha\beta.$$

■

5.2.2 Constructing a Quotient

Purpose. *Given line segments of lengths α and $\beta \neq 0$, to construct a line segment of length α/β.*

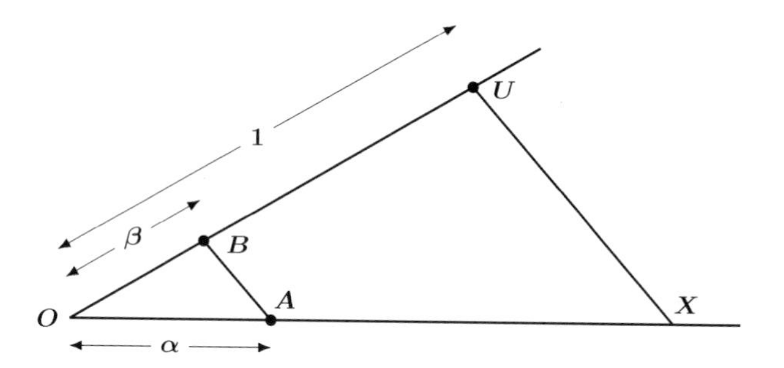

Fig. 5.16 Constructing a quotient

Method. (See Figure 5.16.)

(i), (ii), and (iii) are as in 5.2.1.
(iv) Construct (as in 5.1.7) a line through U parallel to BA and meeting OA (extended, if necessary) at X.

Result. *The line segment OX has length α/β.*

Proof. Let x be the length of OX. By similar triangles,

$$\frac{x}{\alpha} = \frac{1}{\beta}$$

and so

$$x = \frac{\alpha}{\beta}. \qquad \blacksquare$$

5.2.3 Constructing Square Roots

Purpose. *Given a line segment of length $\alpha > 0$, to construct a segment of length $\sqrt{\alpha}$.*

Method. (See Figure 5.17.)

(i) Draw a line segment OA of length α and extend it backwards to U so that OU has length 1.
(ii) Bisect the line segment UA (as in 5.1.1) and construct a semicircle with the midpoint as centre and radius half the length of UA.
(iii) Erect a perpendicular to UA at O (as in 5.1.5) which intersects the semicircle at a point X, say.

Result. *The line segment OX has length $\sqrt{\alpha}$.*

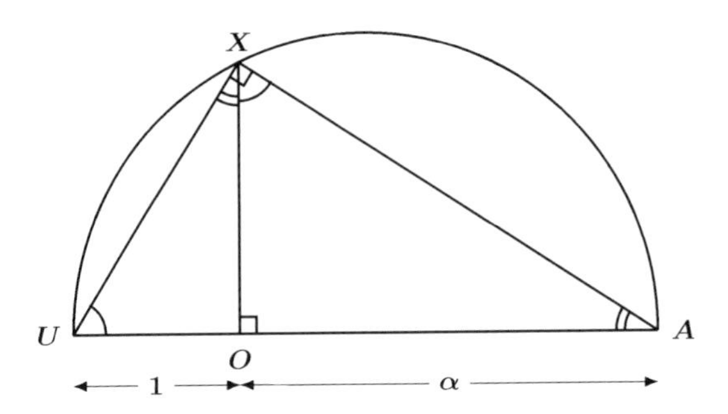

Fig. 5.17 Constructing square roots

Proof. Let x be the length of OX. The angle UXA is a right angle, being the angle subtended by a diameter of the circle at a point on the circumference. Hence the triangles UOX and XOA are similar. It follows that

$$\frac{x}{1} = \frac{\alpha}{x}$$

which gives

$$x = \sqrt{\alpha}.$$

■

Exercises 5.2

1. (a) Start with a line segment which you define to have length 1 (a segment about 2 cm long should be ideal) and construct, with straightedge and compass, line segments of length
   (i) 4, (ii) 1/3, (iii) 7/5.
   (b) Would you now revise your answer to Exercises 5.1 #2(b)?
   (c) For which positive rational numbers r can you construct (with straightedge and compass) a line segment of length r?
2. Start with a line segment you define to have length 1 and then construct, with straightedge and compass, segments of length
   (i) $\sqrt{2}$, (ii) $\sqrt{1+\sqrt{2}}$, (iii) $\sqrt{2-\sqrt{2}}$, (iv) $\sqrt[4]{5}$.
3. Assume you are given a line segment of unit length. Convince yourself that you could construct (but do not bother to carry out the actual constructions accurately) line segments of lengths
   (i) $\sqrt{3/7+6\sqrt{5}}$, (ii) $3\sqrt[4]{7/3}+15/7$.
4. In Figure 5.18 $|OA| = |OB| = 1$, angle $AOB = 36°$, r is the length of AB and the angles PAO and PAB are equal.

   (a) Assume that you are given line segments of lengths 1 and r. Show how you would use straightedge and compass to construct an angle of 36°.

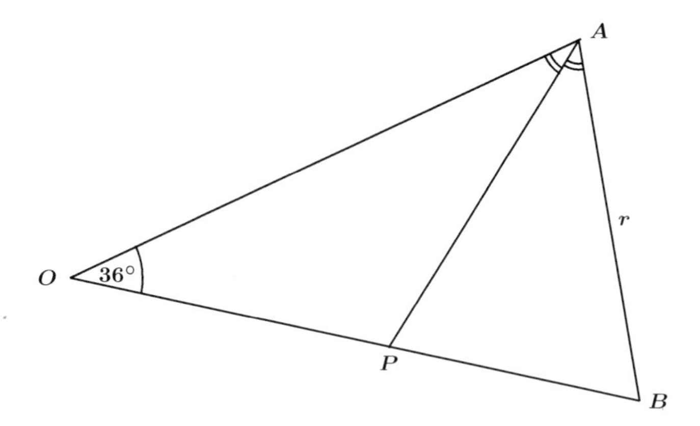

Fig. 5.18 Constructing an angle of 36°

(b) By calculating the sizes of several angles in the figure, show that triangles OAB and APB are similar (isosceles) triangle and hence calculate r.
(c) Starting from a suitable line segment of unit length, construct (with straightedge and compass) a segment of length r.

5. Use Exercise 4 to construct (with straightedge and compass only) a regular pentagon (a figure with 5 equal sides and angles) inscribed inside a circle. (Make the radius of the circle about 4 cm.)

5.3 Rules for Straightedge and Compass Constructions

A fairly precise description of the way the straightedge and compass can (and cannot) be used in performing constructions has already been given in the Introduction and, in more detail, at the start of Section 5.1. The purpose of this section is to spell out in complete detail the very precise rules (as passed down from the Greek geometers) which must be followed in these constructions.

In each of these construction problems, a set of points is given at the start, say

$$\{P_0, P_1, \ldots, P_m\},$$

which we shall call the *initial set* of points for the construction. Some lines and circles may also be given. Finitely many new points

$$P_{m+1}, P_{m+2}, \ldots, P_{m+n}$$

(and extra lines and circles) may be added, using straightedge and compass, to achieve the desired result.

We are not permitted to put points, lines, and circles randomly on the page—rather we must base new lines and circles on points given at the start or already constructed. New points are where these lines and circles meet other lines and circles.

Thus the precise rules are that, at any stage of the construction process, we may add new points (and lines and circles) to the old ones already in existence by performing an operation of one of the following two types:

5.3.1 Construction Rules

(i) *Draw a line through two old points* P_i *and* P_j *to get new points where this line, extended if necessary, intersects other lines and circles.* (See Figure 5.19.)

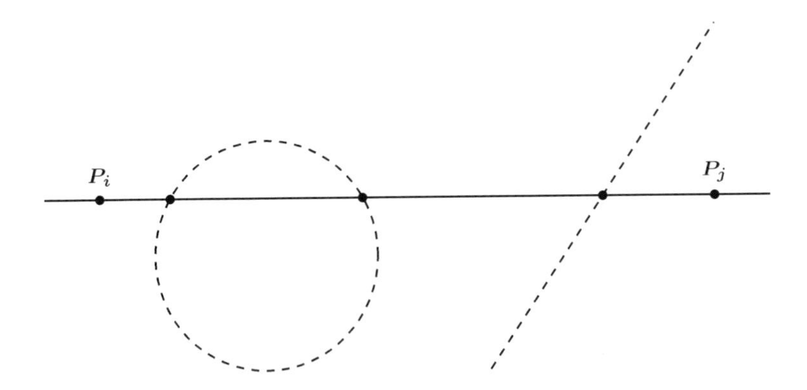

Fig. 5.19 Drawing a line through two given points to intersect other lines and circles

(ii) *Draw a circle with centre at an old point* P_i *and radius equal to the distance between two old points* P_j *and* P_k *to get new points where this circle intersects other lines and circles.* (See Figure 5.20.)

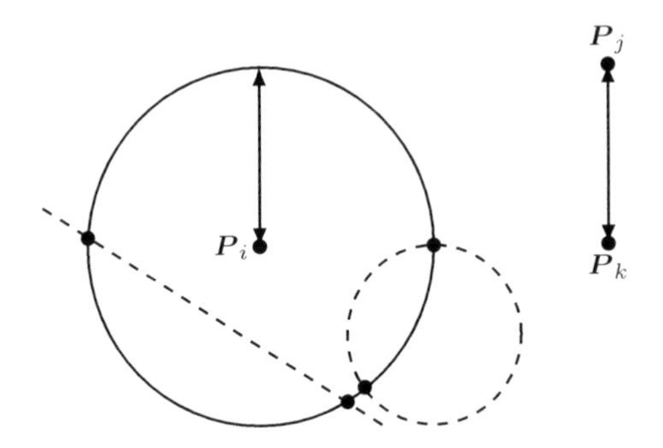

■

Fig. 5.20 Constructing a circle with given centre and radius

Rule (i) tells us very precisely how the straightedge is to be used, while rule (ii) does the same for the compass. These rules tell us the only ways we are allowed to enlarge the set of points, lines and circles.

Note that as a special case, P_j may be the same as P_i in rule (ii). This permits us to draw a circle with centre at an old point P_i and radius equal to the distance between P_i and another old point P_k. (See Figure 5.21.)

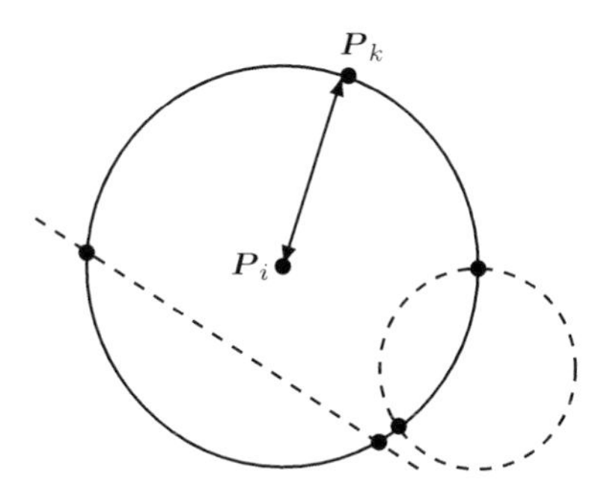

Fig. 5.21 Constructing a circle with given centre and radius equal to the distance between an old point and a new point

Thus rule (ii) does allow us to use the compass in the two ways 2(a) and 2(b) described at the start of Section 5.1. Case 2(a) is when $P_i = P_j$ and 2(b) is when P_i and P_j are different.

In summary then, in any construction problem,

> begin with an initial set of points (and lines and circles), and enlarge the set of points using finitely many operations of types (i) and (ii)

to achieve the desired result.

As you would expect, all of the constructions in Sections 5.1 and 5.2 have been carried out according to these rules. In Example 5.3.2 below, we spell this out for the construction in 5.1.1 and leave you (see Exercises 5.3 # 1–3) to check this for the rest of the constructions.

5.3.2 Example. The construction "bisect a line segment" given in 5.1.1 follows the rules just given.

Proof. The initial set of points for the construction is $\{A, B\}$. Hence, in the notation above we may put $P_0 = A$ and $P_1 = B$.

The construction proceeds by the drawing of two circles with centres A and B respectively, and radius equal to the distance between the points A and B. Thus these circles are of the type permitted by rule (ii), and so the new points D and E, being points where the circles intersect, can be obtained by two operations of type (ii). Hence we may put $P_2 = D$ and $P_3 = E$.

The final stage in the construction involves drawing the line DE joining these two points to get the point C where this line intersects the line AB. Hence C can be obtained by an operation of type (i), and so we may choose $P_4 = C$. The desired mid-point C has thus been obtained from the set of initial points by successive operations of the types (i) and (ii). ■

We conclude this section by re-examining the famous geometric construction problems in the light of the construction rules given above.

5.3.2 *Doubling the Cube*

We are given a cube and asked to construct a cube with double the volume. We first translate this into the following equivalent problem in two dimensions. Given two points P_0 and P_1 whose distance apart is equal to one side of the original cube, we are asked to construct two points P_i and P_j whose distance apart is exactly $\sqrt[3]{2}$ times the distance between P_0 and P_1. (See Figure 5.22.)

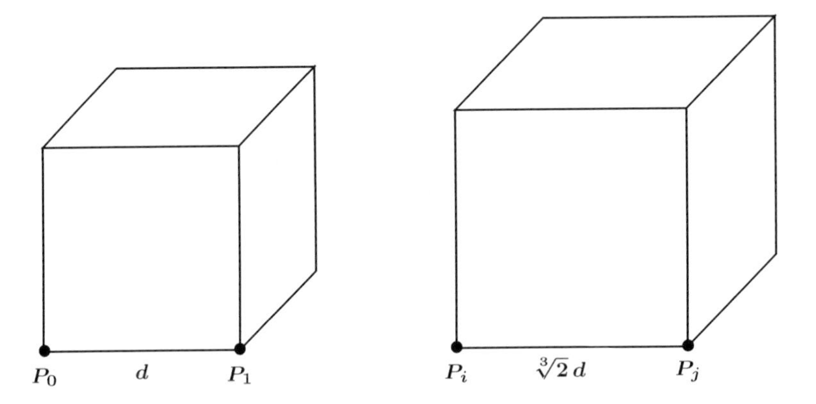

Fig. 5.22 Doubling the cube

The question is whether or not this is possible using operations of types (i) and (ii) only. ■

5.3.3 *Squaring the Circle*

As in 5.3.2, this is equivalent to being given just two points P_0 and P_1 (distance apart equal to the radius of the circle) and being asked to construct two points P_i and

P_j whose distance apart is exactly $\sqrt{\pi}$ times the distance between P_0 and P_1. (See Figure 5.23.)

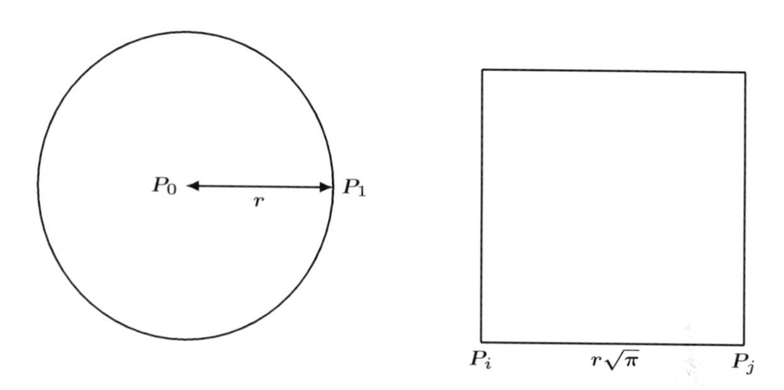

Fig. 5.23 Squaring the circle

As usual, the question is whether or not this is possible using only finitely many operations of types (i) and (ii). ■

5.3.4 *Trisecting an Angle*

We are given the points P_0, P_1 and P_2, which determine the angle. We are then asked to construct (using only operations of types (i) and (ii)) a point P_i such that angle $P_iP_0P_1$ is exactly one third of angle $P_2P_0P_1$. (See Figure 5.24.)

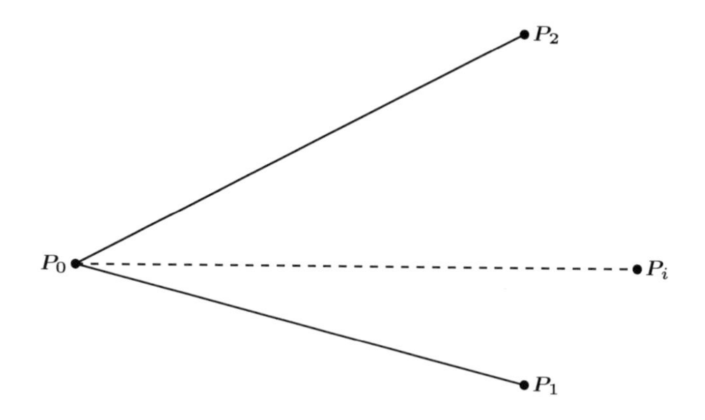

Fig. 5.24 Trisecting an angle

■

Exercises 5.3

1. Verify that the construction 5.1.2 (transferring a length) can be performed by successive operations of types (i) and (ii) of 5.3.1.
2. Verify that the construction 5.1.3 (bisection of an angle) can be performed by successive operations of the types (i) and (ii).
3. Satisfy yourself that all the constructions in Sections 5.1 and 5.2 can be performed by successive operations of types (i) and (ii).

5.4 Constructible Numbers and Fields

In analysing construction problems, it often helps to introduce a convenient unit of length. For example, in analysing the problem of doubling the cube, a convenient unit of length is the length of one side of the given cube. Doubling the cube (see 5.3.2 above) is then possible if, given points P_0 md P_1 one unit apart, we can construct points P_i and P_j whose distance apart is $\sqrt[3]{2}$, using only operations of types (i) and (ii). This leads to the following definition.

5.4.1 Definition. Let γ be a real number with absolute value $|\gamma|$. Then γ is said to be *constructible* if we can construct points P_i and P_j whose distance apart is $|\gamma|$ units by: starting from an initial set of points $\{P_0, P_1\}$ whose distance apart is 1 unit and then performing a finite number of operations of types (i) and (ii) of 5.3.1. ■

Clearly, doubling the cube is possible if and only if the number $\sqrt[3]{2}$ is constructible, while squaring the circle is possible if and only if $\sqrt{\pi}$ is constructible.

The following two theorems and the related discussion make it clear that the set of constructible numbers is quite large. They also mark the place where geometry and algebra begin to overlap.

5.4.2 Theorem. [Field with Square Roots Theorem] *The set CON of all constructible numbers is a subfield of* $\mathbb{R}$. *Furthermore*

(a) all rational numbers are in CON, and
(b) if $\alpha \in CON$ *and* $\alpha > 0$ *then* $\sqrt{\alpha} \in CON$.

Proof.

> We aim to show that *CON* is a subfield of $\mathbb{R}$.
> To do this we shall show that the operations of addition, subtraction, multiplication and division (except by 0) can be performed without restriction in *CON*.

Let α and β be in *CON*. Hence line segments of lengths $|\alpha|$ and $|\beta|$ can be constructed by successive operations of types (i) and (ii) of 5.3.1, starting from a segment of length 1 unit.

Because we can transfer lengths (by 5.1.2), it is easy to see that segments of lengths $|\alpha + \beta|$ and $|\alpha - \beta|$ can be constructed from the above two segments. (If, for example, the segment AB has length α and segment CD has length β, we can extend AB and use 5.1.2 to construct a point E so that BE has length β and AE has the desired length $\alpha + \beta$.) We can also construct, using operations of types (i) and (ii), segments of lengths $|\alpha\beta|$ and $|\alpha/\beta|$ (if $\beta \neq 0$), by 5.2.1 and 5.2.2.

Thus the numbers $\alpha + \beta$, $\alpha - \beta$, $\alpha\beta$ and α/β (if $\beta \neq 0$) are all constructible and so are in *CON*. Hence *CON* is a field.

We shall prove that all rational numbers are in *CON*.
We shall firstly show that all positive integers are in *CON*.
Then we shall use this and 5.2.2, to see that all rational numbers are in *CON*.

Since we are given a segment of length 1, we can construct segments of lengths $1 + 1 = 2, 2 + 1 = 3$ and so on, from which we see that we can construct segments of length equal to any positive integer. Hence, by 5.2.2, we can construct a segment of length equal to any desired positive rational number m/n (with $m, n \in \mathbb{N}$). Because of the use of $|\gamma|$ in Definition 5.4.1, it follows that all rational numbers (negative as well as positive) are in *CON*. (Alternatively, the result that $\mathbb{Q} \subseteq CON$ follows from Proposition 1.1.1 and the fact that *CON* is a field.)

Finally, by 5.2.3, if $\alpha \in CON$ and $\alpha > 0$, then $\sqrt{\alpha}$ is constructible. ■

By applying this theorem repeatedly we see that each of the numbers

$$1 + 1 = 2, \quad 5, \quad \frac{2}{5}, \quad \frac{2}{5} + \sqrt{2}, \quad 3 + \sqrt{\frac{2}{5} + \sqrt{2}}, \quad 2 + \sqrt{3 + \sqrt{\frac{2}{5} + \sqrt{2}}}$$

is constructible.

Thus we see that every real number which is obtained from $\mathbb{Q}$ by performing successive field operations and taking square roots is constructible. The following theorem expresses this precisely, in the notation of field extensions.

5.4.3 Theorem. [Successive Square Roots Give Constructibles] *A real number γ is constructible if there exist positive real numbers $\gamma_1, \gamma_2, \ldots, \gamma_n$ such that*

$\gamma_1 \in \mathbb{F}_1$ *where* $\mathbb{F}_1 = \mathbb{Q}$,
$\gamma_2 \in \mathbb{F}_2$ *where* $\mathbb{F}_2 = \mathbb{F}_1(\sqrt{\gamma_1})$,
$\vdots \qquad \vdots$
$\gamma_n \in \mathbb{F}_n$ *where* $\mathbb{F}_n = \mathbb{F}_{n-1}(\sqrt{\gamma_{n-1}})$,
and, finally,
$\gamma \in \mathbb{F}_{n+1}$ *where* $\mathbb{F}_{n+1} = \mathbb{F}_n(\sqrt{\gamma_n})$.

Proof. This follows immediately from the Field with Square Roots Theorem 5.4.2. ■

Notice that there is a tower of fields

$$\mathbb{Q} \subseteq \mathbb{F}_1 \subseteq \mathbb{F}_2 \subseteq \ldots \subseteq \mathbb{F}_n \subseteq \mathbb{F}_{n+1}$$

implicit in this theorem. This tower will play an important rôle in the impossibility proofs in the next chapter.

In the following example, the fields $\mathbb{F}_1, \mathbb{F}_2, \ldots$ and the numbers $\gamma_1, \gamma_2, \ldots$ relevant to a particular γ are indicated.

5.4.4 Example. Let

$$\gamma = 5\sqrt{2} + \frac{\sqrt{8 - 3\sqrt{2}}}{(1 - \sqrt{2})}.$$

Prove that γ is constructible by using the Successive Square Roots Give Constructibles Theorem 5.4.3.

Proof.

> We must produce a positive integer n and positive real numbers $\gamma_1, \gamma_2, \ldots, \gamma_n$ as in the statement of the Successive Square Roots Give Constructibles Theorem 5.4.3 such that
>
> $$\gamma \in \mathbb{F}_{n+1} = \mathbb{F}_n(\sqrt{\gamma_n}).$$

$$\begin{aligned} \text{Put} \quad && \gamma_1 &= 2 && \in \mathbb{F}_1 \quad \text{where} \quad \mathbb{F}_1 = \mathbb{Q}, \\ && \gamma_2 &= 8 - 3\sqrt{2} && \\ && &= 8 - 3\sqrt{\gamma_1} && \in \mathbb{F}_2 \quad \text{where} \quad \mathbb{F}_2 = \mathbb{F}_1(\sqrt{\gamma_1}). \\ \text{Hence} \quad && \gamma &= 5\sqrt{\gamma_1} + \frac{\sqrt{\gamma_2}}{1 - \sqrt{\gamma_1}} && \in \mathbb{F}_3 \quad \text{where} \quad \mathbb{F}_3 = \mathbb{F}_2(\sqrt{\gamma_2}). \end{aligned}$$

Thus we have produced positive real numbers γ_1, γ_2 which satisfy the hypothesis of the Successive Square Roots Give Constructibles Theorem 5.4.3. Hence γ is constructible. ■

In this chapter we have concentrated on showing that there are many different constructible numbers. In the next chapter (where we prove the impossibility of the three famous constructions) we take the opposite point of view and concentrate on showing that there are many real numbers which are not constructible. Indeed, we shall prove there that the only constructible numbers are those given by the

Successive Square Roots Give Constructibles Theorem 5.4.3.

That is, we prove that any number which cannot be obtained from $\mathbb{Q}$ by performing successive field operations and taking square roots is not constructible. This result, which we call the

All Constructibles Come From Square Roots Theorem,

is the converse of Theorem 5.4.3.

Exercises 5.4

1. Let $\gamma = 3\sqrt{2} + \sqrt{5 - 3\sqrt{2}}$.
   Prove in the following two different ways that γ is constructible:
   (a) firstly, by applying the Field with Square Roots Theorem 5.4.2,
   (b) secondly, by applying the Successive Square Roots Give Constructibles Theorem 5.4.3.
2. Show that the zeros in $\mathbb{R}$ of the polynomial $2X^4 - 6X^2 - 3$ are all constructible real numbers.
   [Hint. This is a quadratic in X^2.]
3. Given constructible numbers α and β, describe how you would use 5.1.2 to construct a segment of length $|\alpha + \beta|$ in each of the following cases:
   (i) α and β have the same sign,
   (ii) α and β have opposite sign.
4. [This exercise shows that, in some sense, if we could use infinitely many operations of types (i) and (ii), every real number would be constructible.]
   (a) Show that every real number can be approximated arbitrarily closely by a rational number.
   (b) Show that every real number can be approximated arbitrarily closely by a constructible number.

Additional Reading for Chapter 5

Basic straightedge and compass constructions, including bisection of line segments and angles, are given in Chapter 4, (Jacobs, 1953, Lesson 8); proofs that the constructions achieve their intended purpose are included. A discussion of collapsing and modern (or noncollapsing) compasses is given in (Eves, 1972, §4.1). The constructions for sums, differences, products, quotients, and square roots of constructible numbers are given in (Fraleigh, 1982, §39.1). The rules governing the use of straightedge and compass are explained in (Hobson, 1953, pp. 7–8).

Those who have previously met the constructions of bisecting a line segment, bisecting an angle, etc, will notice that our constructions do not use arbitrary choices.

Phrases which are common in elementary geometry textbooks such as "draw a *suitable* arc" or "choose a *suitable* radius" are absent from our presentation. For a discussion of such indeterminate constructions see (Hobson, 1953, p. 13).

It is interesting to note that all constructions that can be performed using straightedge and compass can also be performed using a compass alone. This was first proved in Mascheroni (1797) by the Italian mathematician Lorenzo Mascheroni (1750–1800). (Mascheroni is also known for the Euler-Mascheroni constant γ.) It is also true that all such constructions can be performed using a straightedge and a single circle, or even an arc of a circle, together with its centre. For a discussion of this work of Jean Victor Poncelet (1788–1867), Jacob Steiner (1796–1863), Lorenzo Mascheroni (1750–1800), and Georg Mohr (1640–1697) see (Gardner, 1981, Chapter 17), and also Kostovskii (1961) and Steiner (1948).

A very different problem from that considered in this book is that of squaring the circle using scissors. More precisely, the problem is to cut a circle (or rather, a disc) into finitely many pieces using scissors (cutting along Jordan arcs) and then to reassemble the pieces into a square of the same area. It is proved in Dubins et al. (1963), however, that this construction is impossible. Whether the construction is possible when more weird "pieces" are allowed than those obtained by cutting with scissors was an unsolved problem until 1990 when the Hungarian mathematician Miklós Laczkovich (born 1948) in Laczkovich (1990) proved the surprising result that it can be done. He used about 10^{50} pieces. An exposition in an undergraduate journal appears in Pearce et al. (2009) which can be downloaded. More recently Marks and Unger (2017) gave a completely constructive solution to Tarski's circle squaring problem. The analogous problem in three (or higher) dimensions is solved by the Banach-Tarski (Paradox) Theorem. An exposition of this astonishing theorem appears in French (1988).

Chapter 6
Proofs of the Geometric Impossibilities

Within this chapter it finally becomes clear why the three famous geometrical constructions are impossible. We are now, at long last, in a position to see the solutions of problems which defied the world's best mathematicians for over two thousand years. The key to the solutions lies in combining the geometrical ideas from Chapter 5 with the algebraic ideas from earlier chapters.

Our first task is to show that there are no constructible numbers other than those already found in Chapter 5. This enables us to show that every constructible number is algebraic over $\mathbb{Q}$ and has a degree which is a power of 2.

We are then able to show that, if doubling the cube or trisecting an arbitrary angle were possible, there would have to exist a constructible number whose degree is not a power of 2. This contradiction means that we can be certain that these constructions (Problems I and II) are impossible.

We complete the proof of the impossibility of squaring the circle (Problem III) by showing in Theorem 10.5.6 of Chapter 10 that π is not algebraic over $\mathbb{Q}$.

6.1 Non-Constructible Numbers

In Chapter 5 we found that real numbers like

$$2+\sqrt{3+\sqrt{\frac{2}{5}}+\sqrt{2}},$$

which are obtained from elements of $\mathbb{Q}$ by successively applying field operations and taking square roots, are all constructible. For the purposes of this chapter, however, it is of the utmost importance to know whether we can go in the reverse direction:

S. A. Morris et al., *Abstract Algebra and Famous Impossibilities*, Undergraduate Texts in Mathematics, https://doi.org/10.1007/978-3-031-05698-7_6

can every constructible number be expressed in terms of repeated square roots and field operations, starting from numbers in $\mathbb{Q}$?

The following theorem shows that the answer is "yes".

6.1.1 Theorem. [All Constructibles Come From Square Roots] *If the real number γ is constructible, then there exist positive real numbers $\gamma_1, \gamma_2, \ldots, \gamma_n$ such that*

$\gamma_1 \in \mathbb{F}_1$ *where* $\mathbb{F}_1 = \mathbb{Q}$,
$\gamma_2 \in \mathbb{F}_2$ *where* $\mathbb{F}_2 = \mathbb{F}_1(\sqrt{\gamma_1})$,
$\vdots \qquad \vdots$
$\gamma_n \in \mathbb{F}_n$ *where* $\mathbb{F}_n = \mathbb{F}_{n-1}(\sqrt{\gamma_{n-1}})$,
and, finally,
$\gamma \in \mathbb{F}_{n+1}$ *where* $\mathbb{F}_{n+1} = \mathbb{F}_n(\sqrt{\gamma_n})$.

Proof. We postpone this rather long proof until Section 6.3. The idea of the proof is that when you intersect lines and circles, the worst that can happen is that you need extra square roots to describe the coordinates of the points of intersection. The example below shows this happening in a particular case. ■

6.1.2 Example. Start from two points $P_0 = (0, 0)$ and $P_1 = (1, 0)$, at a distance 1 apart, as in the definition of a constructible number (Definition 5.4.1). Draw two circles with centres at P_0 and P_1 respectively, each with radius P_0P_1. Show that the coordinates of each of the points of intersection, P_2 and P_3, of the two circles involve a single square root. (See Figure 6.1.)

Proof. The circles have equations

$$x^2 + y^2 = 1, \tag{1}$$

$$(x - 1)^2 + y^2 = 1. \tag{2}$$

The points of intersection satisfy both equations. Subtracting (1) from (2) gives $x = 1/2$ and substituting in (1) gives $y = \pm\sqrt{3}/2$. Thus the two points of intersection are $P_2 = \left(\frac{1}{2}, \frac{\sqrt{3}}{2}\right)$ and $P_3 = \left(\frac{1}{2}, -\frac{\sqrt{3}}{2}\right)$, which involve nothing worse than a square root, $\sqrt{3}$. ■

We now apply the All Constructibles Come From Square Roots Theorem 6.1.1 to show that every constructible number must be algebraic over $\mathbb{Q}$ and must have degree over $\mathbb{Q}$ which is a power of 2. This result, which is the key to the impossibility proofs, enables us to be certain that many numbers are tnot constructible.

6.1.3 Theorem. [Degree of a Constructible Number Theorem] *If a real number γ is constructible, then γ is algebraic over $\mathbb{Q}$ and* $\deg(\gamma, \mathbb{Q}) = 2^s$ *for some integer $s \geq 0$.*

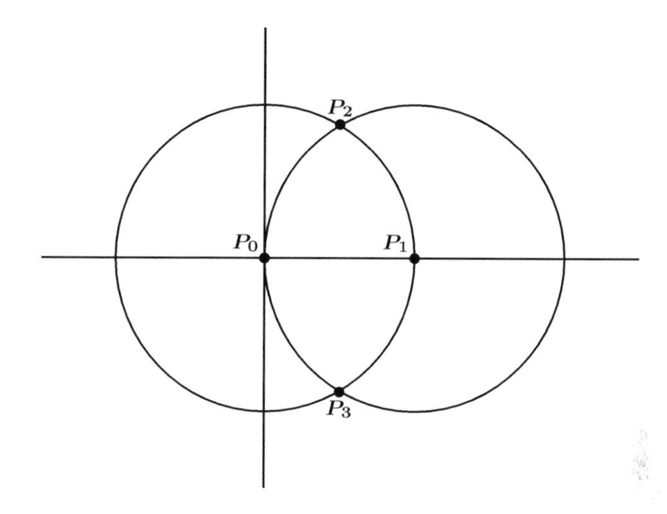

Fig. 6.1 Intersection of two circles

Proof. Let γ be constructible and let $\gamma_1, \ldots, \gamma_n$ be as in the "All Constructibles Come From Square Roots" Theorem. The number $\sqrt{\gamma_i}$ is a zero of the polynomial $X^2 - \gamma_i$, which is in $\mathbb{F}_i[X]$ since $\gamma_i \in \mathbb{F}_i$. Hence, by Definition 2.3.2,

$$\deg(\sqrt{\gamma_i}, \mathbb{F}_i) = 1 \text{ or } 2$$

and since $\mathbb{F}_{i+1} = \mathbb{F}_i(\sqrt{\gamma_i})$ it follows from Theorem 3.2.4 that

$$[\mathbb{F}_{i+1} : \mathbb{F}_i] = 1 \text{ or } 2 \quad (1 \leq i \leq n).$$

Repeated application of the Dimension for a Tower Theorem 3.4.3 to the tower

$$\mathbb{Q} = \mathbb{F}_1 \subseteq \mathbb{F}_2 \subseteq \mathbb{F}_3 \subseteq \ldots \subseteq \mathbb{F}_{n+1}$$

shows that

$$\begin{aligned}[\mathbb{F}_{n+1} : \mathbb{Q}] &= [\mathbb{F}_{n+1} : \mathbb{F}_n][\mathbb{F}_n : \mathbb{F}_{n-1}] \ldots [\mathbb{F}_2 : \mathbb{F}_1] \\ &= 2^u, \quad \text{for some integer } u \geq 0.\end{aligned}$$

It follows from Theorem 4.4.1 that γ is algebraic over $\mathbb{Q}$. Also, by considering the tower

$$\mathbb{Q} \subseteq \mathbb{Q}(\gamma) \subseteq \mathbb{F}_{n+1}$$

we see that $\deg(\gamma, \mathbb{Q})$ is a factor of $[\mathbb{F}_{n+1} : \mathbb{Q}]$. Hence

$$\deg(\gamma, \mathbb{Q}) = 2^s$$

for some integer $s \geq 0$. ■

It should be noted, perhaps with a little surprise, that the *converse* of the above result is false. A counterexample to the converse is given as (Gilbert, 1976, Example 13–18); this example describes a specific number γ which is not constructible but which is algebraic over $\mathbb{Q}$ with $\deg(\gamma, \mathbb{Q}) = 2^2$.

Exercises 6.1

1. Let $\gamma = 2 + \sqrt{3 + \sqrt{5}}$.
   (a) Prove that γ is constructible by using the Field with Square Roots Theorem 5.4.2.
   (b) Prove that γ is constructible by using the Successive Square Roots Give Constructibles Theorem 5.4.3.
   [Hint. Show how to choose γ_1, γ_2, etc.]
   (c) What does the Degree of a Constructible Number Theorem 6.1.3 now tell you about the degree of γ?
2. Combine the Successive Square Roots Give Constructibles and the All Constructibles Come From Square Roots theorems into one theorem.
3. Use the Degree of a Constructible Number Theorem 6.1.3 and the fact that

$$\operatorname{irr}(\sqrt[3]{5}, \mathbb{Q}) = X^3 - 5$$

   to prove that $\sqrt[3]{5}$ is not constructible.
4. (a) Use the Degree of a Constructible Number Theorem to prove that the number $1 + \sqrt[3]{5}$ is not constructible. Include a full proof that $\deg(1 + \sqrt[3]{5}, \mathbb{Q})$ is what you claim it to be.
   (b) Use the result of Exercise 3 and the fact that the constructible numbers form a field to give an alternative proof that the number $1 + \sqrt[3]{5}$ is not constructible.
5. In Example 6.1.2 let P_4 be the point of intersection of the line segments P_0P_1 and P_2P_3.
   (a) Write down the equation of the circle with centre P_2 and radius P_2P_4.
   (b) Show that this new circle intersects the two circles already constructed in four points all of whose coordinates lie in the field $\mathbb{Q}(\sqrt{3})(\sqrt{39})$.

6.** Show that the the set of all constructible numbers is countably infinite.
[Hint. Recall that a set S is said to be *countably infinite* if there exists a function $f : \mathbb{N} \to S$ which is one-to-one and onto. See Wilder (2012) and Halmos (1960)]

6.2 The Three Geometric Constructions are Impossible

At long last, we are in a position to see why the constructions in the Introduction are impossible. In each case we shall show that if the construction were possible, then there would exist a constructible number which is not algebraic or whose degree over $\mathbb{Q}$ is not a power of 2. Because the existence of such a constructible number would contradict the Degree of a Constructible Number Theorem 6.1.3, we can be certain that each construction is impossible.

6.2.1 Problem I - Doubling the cube.

A cube whose volume is 2 must have sides whose lengths are $\sqrt[3]{2}$. Hence doubling the cube amounts to constructing a line segment whose length is $\sqrt[3]{2}$ (starting from a line segment of length 1 and using only straightedge and compass). Thus if the cube could be doubled, then $\sqrt[3]{2}$ would be a constructible number.

By Example 4.3.3, however,

$$\operatorname{irr}(\sqrt[3]{2}, \mathbb{Q}) = X^3 - 2$$

and so

$$\deg(\sqrt[3]{2}, \mathbb{Q}) = 3,$$

which is not a power of 2. This shows that $\sqrt[3]{2}$ is not a constructible number, and so the cube cannot be doubled.

6.2.2 Problem II - Trisecting an arbitrary angle.

If we could trisect every angle then, in particular, we could trisect an angle of 60°. But since 60° is an angle which can be constructed (see Section 5.1), it follows that we could then construct an angle of 20° also. Hence, using a right-angled triangle (see Exercises 6.2 #2 below), we could also construct a line segment of length cos(20°).

It is easy to show, however, that the irreducible polynomial of cos(20°) over $\mathbb{Q}$ is a cubic (see Exercises 6.2 #3 below for the details). Hence

$$\deg(\cos(20^\circ), \mathbb{Q}) = 3$$

and so cos(20°) is not constructible, as 3 is not a power of 2. Thus it is not possible to construct an angle of 20°.

You should note that the argument just given shows that there is one angle (namely 60°) which cannot be trisected. But it is easy to deduce from this result (as, for example, in Exercises 6.2 #4 below) that there are infinitely many angles which cannot be trisected.

Note also that the argument given above that the angle 60° cannot be trisected, relies crucially on the fact that we can construct an angle of 60°. Indeed, the argument given cannot be used to decide whether it is possible to trisect other angles (such as 10°) which cannot themselves be constructed (starting just from a pair of points P_0 and P_1).

6.2.3 *Problem III - Squaring the circle.*

A circle of radius 1 has area equal to π square units, and so a square with the same area would have sides of length $\sqrt{\pi}$. Thus if squaring the circle could be done with straightedge and compass, it would follow that $\sqrt{\pi}$ is constructible. This would mean (by Theorem 6.1.3) that $\sqrt{\pi}$ is algebraic over $\mathbb{Q}$, and hence that $\pi = \sqrt{\pi}.\sqrt{\pi}$ is algebraic over $\mathbb{Q}$ by Theorem 4.4.4. We complete the proof of the impossibility of squaring the circle in Theorem 10.5.6 of Chapter 10 in which we show that π is transcendental (that is, not algebraic over $\mathbb{Q}$).

6.2.4 *Other constructions which are impossible.*

In Hadlock (1978) it is shown that not every angle can be divided into *five* equal parts, with straightedge and compass. As a generalization, we indicate a proof below (in Exercises 6.2 #7 and 8) that, for any given positive integer n which is not a power of 2, it is impossible to divide an arbitrary angle into n equal parts.

Although there are proofs for the impossibility of trisecting angles which do not involve the full algebraic machinery developed in this book, it is remarked in Hadlock (1978) that such proofs are not so easy to adapt to proving further impossibilities such as that of quintisecting an arbitrary angle.

Exercises 6.2

1. Decide in each case whether the number is constructible and give reasons. You may assume $\sqrt[3]{2}$ is not constructible.

   (i) $\sqrt{2+\sqrt{2}}$, (ii) $\sqrt[4]{2}$, (iii) $\sqrt[6]{2}$,
   (iv) $\sqrt{2}\sqrt[3]{2}$, (v) $\sqrt{2}+\sqrt[3]{2}$.

2. (a) Assume that an angle θ has been constructed. Show how you would construct lengths $\sin\theta$ and $\cos\theta$ from it, using straightedge and compass.
   (b) Assume, conversely, that a length $\sin\theta$ has been constructed.
      (i) Show *algebraically* that $\cos\theta$ is then constructible.
      [Hint. Use the Field with Square Roots Theorem 5.4.2.]
      (ii) Show *geometrically* how to construct $\cos\theta$.

(iii) Show how you would use straightedge and compass to construct the angle θ $(0 < \theta < \frac{\pi}{2})$ from $\sin\theta$ and $\cos\theta$.

3. (a) Use the identity $\cos(3\theta) = -3\cos\theta + 4\cos^3\theta$ to show that $\cos(20°)$ is a zero of the polynomial $8X^3 - 6X - 1$.
   (b) Deduce that $\cos(20°)$ is not a constructible number. Which part of Exercise 2 now shows that an angle of $20°$ cannot be constructed with straightedge and compass?
   (c) Which angle have you proved cannot be trisected?

4. (a) Show that an angle of $30°$ cannot be trisected. [Hint. Note that an angle of $60°$ cannot be trisected.]
   (b) Repeat for an angle of $15°$.
   (c) Show that *there are infinitely many angles which cannot be trisected.*

5. (a) Let $\alpha \in \mathbb{R}$ be a zero of the polynomial $p(X) = X^3 + 2X + 1$ in $\mathbb{Q}[X]$. Prove each of the following:
      (i) $p(X)$ is irreducible over $\mathbb{Q}$.
      (ii) $p(X) = \mathrm{irr}(\alpha, \mathbb{Q})$.
      (iii) α is not constructible.
   (b) Can we assert that $1 - \alpha + \alpha^2 \neq 0$? Why?

6. Let α and $\beta \in \mathbb{R}$. In each case decide whether the sentence is true or false. Give reasons.

   (a) If α and β are constructible then so is $\alpha + \beta$.
   (b) If $\alpha + \beta$ is constructible then so are α and β.
   (c) If $\alpha\beta$ is constructible then so are α and β.
   (d) If $\alpha + \beta$ and $\alpha\beta$ are both constructible then so are α and β. [Hint for (d). Can you express α and β in terms of $\alpha + \beta$ and $\alpha\beta$?]

7.* In this exercise you will prove the Cos $(n\theta)$-formula: *For each prime number $n > 2$ there are integers $a_1, a_3, \ldots, a_{n-2}, a_n$ such that*

$$\cos(n\theta) = a_1\cos\theta + a_3\cos^3\theta + \ldots + a_{n-2}\cos^{n-2}\theta + a_n\cos^n\theta$$

for all $\theta \in \mathbb{R}$, where $a_n = 2^{n-1}$ and the remaining coefficients $a_1, a_3, \ldots, a_{n-2}$ all have n as a factor.

   (a) Using the De Moivre Theorem (see Ledermann (1965)) and the Binomial Theorem show that, if we let $c = \cos\theta$, then

$$\cos(n\theta) = c^n + \binom{n}{2}c^{n-2}(c^2 - 1) + \binom{n}{4}c^{n-4}(c^2 - 1)^2 + \ldots + \binom{n}{n-1}c(c^2 - 1)^{\frac{n-1}{2}}.$$

   The following parts of this exercise assume that the right hand side of this formula is rearranged to give a polynomial in c.

(b) Show that the coefficient of c^n in this polynomial is

$$1 + \binom{n}{2} + \binom{n}{4} + \ldots + \binom{n}{n-1}$$

and then use the Binomial Theorem to show that this equals

$$\frac{1}{2}((1+1)^n + (1+(-1))^n) = 2^{n-1}.$$

(c) By evaluating both sides of the equation in (a) at suitable θ, show that the constant term of the polynomial is zero.

(d) State why the prime number n is a factor of $\binom{n}{r}$ for $1 \le r \le n-1$. Deduce that the coefficients of the remaining powers of c in the polynomial in part (a) all have n as a factor.

8.* **Eisenstein's Irreducibility Criterion** states: *the polynomial*

$$a_0 + a_1 X + \ldots + a_{n-1} X^{n-1} + a_n X^n \in \mathbb{Z}[X]$$

is irreducible over $\mathbb{Q}$ *if there exists a prime number* p *such that* p *is not a factor of* a_n, *but* p is *a factor of* $a_0, \ldots, a_{n-1}$, *and* p^2 *is not a factor of* a_0. (For a proof see Corollary 9.2.2 or (Fraleigh, 1982, Theorem 31.4).)

(a) Let $n > 2$ be a prime number. Choose θ so that $\cos(n\theta) = \frac{n}{n+1}$ and then deduce from the Cos($n\theta$)-formula and the Eisenstein Irreducibility Criterion that $\deg(\cos\theta, \mathbb{Q}) = n$.

(b) Deduce that, for each positive integer n which is not a power of 2, there exists an angle which cannot be divided into n equal parts via straightedge and compass constructions.

6.3 All Constructibles Come From Square Roots" Theorem

In order to describe the points (and the associated lines and circles) which appear when the operations (i) and (ii) of 5.3.1 are applied, we work with the coordinates of these points. To do so, we introduce the following definitions.

6.3.1 Definitions. Let $\mathbb{F}$ be a subfield of $\mathbb{R}$.
A point is an $\mathbb{F}$-*point* if both of its coordinates are in $\mathbb{F}$.
A line is an $\mathbb{F}$-*line* if it passes through two $\mathbb{F}$-points.
A circle is an $\mathbb{F}$-*circle* if its centre is an $\mathbb{F}$-point and its radius is the distance between two $\mathbb{F}$-points. ■

6.3.2 Example. (a) $(2, 1)$ is a $\mathbb{Q}$-point, because both its coordinates belong to $\mathbb{Q}$.
(b) $(1, \sqrt{2})$ is a $\mathbb{Q}(\sqrt{2})$-point.
(c) $\{(x, y) : y = x\}$ is a $\mathbb{Q}$-line, because it contains the two $\mathbb{Q}$-points $(0, 0)$ and $(1, 1)$.
(d) $\{(x, y) : y = \sqrt{2}x\}$ is a $\mathbb{Q}(\sqrt{2})$-line but not a $\mathbb{Q}$-line.
(e) $\{(x, y) : x^2 + y^2 = 4\}$ is a $\mathbb{Q}$-circle, because it has the $\mathbb{Q}$-point $(0, 0)$ as its centre while its radius is the distance between the $\mathbb{Q}$-points $(0, 0)$ and $(0, 2)$.
(f) $\{(x, y) : x^2 + y^2 = 2\}$ is a $\mathbb{Q}(\sqrt{2})$-circle. ■

If we intersect a $\mathbb{Q}$-line with a $\mathbb{Q}$-circle, do we get a $\mathbb{Q}$-point in every case? The following example shows that the answer is "no".

6.3.3 Example. The $\mathbb{Q}$-line $\{(x, y) : y = x\}$ intersects the $\mathbb{Q}$-circle $\{(x, y) : x^2 + y^2 = 4\}$ at the two points $(\sqrt{2}, \sqrt{2})$ and $(-\sqrt{2}, -\sqrt{2})$. These points are not $\mathbb{Q}$-points. However, they are $\mathbb{Q}(\sqrt{2})$-points. ■

6.3.4 Remarks. Assume that we are performing a construction as described in Section 5.3 and that we have already obtained points $P_0, P_1, \ldots, P_k$.

Let $\mathbb{F}$ be a subfield of $\mathbb{R}$ containing the coordinates of all the points $P_0, P_1, \ldots, P_k$ so that these points are all $\mathbb{F}$-points (from Definition 6.3.1).

If we perform an operation of type (i) or (ii) of 5.3.1, we might obtain several new points $P_{k+1}, \ldots, P_{k+t}$. The lines and circles which appear in (i) and (ii) are $\mathbb{F}$-lines and $\mathbb{F}$-circles, and so each of the new points obtained is at the intersection of

(1) two $\mathbb{F}$-lines,
or (2) an $\mathbb{F}$-line and an $\mathbb{F}$-circle,
or (3) two $\mathbb{F}$-circles.

The next lemma tells us that, at worst, the coordinates of these new points involve extra square roots. ■

6.3.5 Lemma. *Let $\mathbb{F}$ be a subfield of $\mathbb{R}$.*

(a) If two $\mathbb{F}$-lines intersect in a single point, then the point is an $\mathbb{F}$-point.
(b) Given an $\mathbb{F}$-line and an $\mathbb{F}$-circle, there exists a positive number $\alpha \in \mathbb{F}$ such that the points of intersection (if any) of this line and circle are $\mathbb{F}(\sqrt{\alpha})$-points.
(c) Given two $\mathbb{F}$-circles, there exists a positive number $\alpha \in \mathbb{F}$ such that any points of intersection of these two circles are $\mathbb{F}(\sqrt{\alpha})$-points.

Proof. Note that $\mathbb{F}$-lines have equations of the form

$$dx + ey + f = 0 \tag{1}$$

for d, e, f in $\mathbb{F}$, while $\mathbb{F}$-circles have equations of the form

$$x^2 + y^2 + px + qy + r = 0 \tag{2}$$

for p, q, r in $\mathbb{F}$.

(a) To find where two $\mathbb{F}$-lines meet, we solve two simultaneous equations of the form (1). This can be done just using field operations $+, -, ., /$, so the solutions are both in $\mathbb{F}$.

(b) Finding where a line and a circle meet amounts to solving two equations, one of the form (1) and the other like (2). This is easily done by using the

$$\frac{-b \pm \sqrt{b^2 - 4ac}}{2a}$$

formula for solving the quadratic $ax^2 + bx + c = 0$. Since, at worst, square roots are introduced, we obtain $\mathbb{F}(\sqrt{\alpha})$-points for some positive $\alpha \in \mathbb{F}$.

(c) The two $\mathbb{F}$-circles

$$x^2 + y^2 + p_1 x + q_1 y + r_1 = 0,$$
$$x^2 + y^2 + p_2 x + q_2 y + r_2 = 0$$

meet where the first of these and the $\mathbb{F}$-line

$$(p_1 - p_2)x + (q_1 - q_2)y + (r_1 - r_2) = 0$$

meet. So this case follows from (b). ■

With these preliminaries out of the way, we can now prove the All Constructibles Come From Square Roots Theorem 6.1.1.

Proof of Theorem 6.1.1. Assume that γ is constructible so that, by the definition of constructible number (Definition 5.4.1), there exists a set of points

$$P_0, P_1, P_2, \ldots, P_n,$$

constructed as in Section 5.3, with $|\gamma|$ equal to the distance between two of these points. Because the initial points P_0 and P_1 are at a distance 1 unit apart, we can introduce coordinate axes so that $P_0 = (0, 0)$ and $P_1 = (1, 0)$. Note that the coordinates of P_0 and P_1 are in $\mathbb{Q}$.

Initially, the points P_0, P_1 are $\mathbb{Q}$-points. Put $\mathbb{F}_1 = \mathbb{Q}$. Assume that for some k satisfying $1 \leq k \leq n-1$, the points $P_0, P_1, \ldots, P_k$ are $\mathbb{F}_k$-points, where $\mathbb{F}_k$ is a subfield of $\mathbb{R}$.

It follows from the Construction Rules in 5.3.1 that the next point, P_{k+1}, will be at the intersection of

(1) a pair of $\mathbb{F}_k$-lines,
or (2) an $\mathbb{F}_k$-line and an $\mathbb{F}_k$-circle,
or (3) a pair of $\mathbb{F}_k$-circles.

Hence, by Lemma 6.3.5, P_{k+1} is an $\mathbb{F}_{k+1}$-point where

$$\mathbb{F}_{k+1} = \mathbb{F}_k(\sqrt{\gamma_k}), \text{ for some positive } \gamma_k \text{ in } \mathbb{F}_k.$$

Since $\mathbb{F}_k \subseteq \mathbb{F}_{k+1}$, this implies further that

$$P_0, P_1, \ldots, P_k, P_{k+1} \text{ are } \mathbb{F}_{k+1}\text{-points.}$$

From the boxed statements, it now follows by mathematical induction on k that the numbers $\gamma_1, \ldots, \gamma_n$ are as described in Theorem 6.1.1. It also follows that $|\gamma|$, being the distance between two of the points

$$P_0, P_1, \ldots, P_n$$

(which are $\mathbb{F}_n$-points), will be the square root of an element of $\mathbb{F}_n$. Thus, for some $\gamma_n \in \mathbb{F}_n$ with $\gamma_n > 0$,

$$\gamma \in \mathbb{F}_n(\sqrt{\gamma_n}).$$ ■

Exercises 6.3

1. [This illustrates case (a) of the proof of Lemma 6.3.5.] Show that the point where the two $\mathbb{Q}(\sqrt{2})$-lines

$$\left(1+\sqrt{2}\right)x + y + 3 - \sqrt{2} = 0,$$
$$2\sqrt{2}x + \left(4+\sqrt{2}\right)y + 1 = 0$$

meet is a $\mathbb{Q}(\sqrt{2})$-point.

2. [This is the general proof of case (a) of the proof of Lemma 6.3.5.] Consider two $\mathbb{F}$-lines

$$d_1x + e_1y + f_1 = 0 \qquad (1)$$
$$d_2x + e_2y + f_2 = 0 \qquad (2)$$

(with $d_1, d_2, e_1, e_2, f_1, f_2 \in \mathbb{F}$) which are not parallel.

(i) If $e_1 \neq 0$, use (1) to get an expression for y, substitute this into (2) and hence find where the two lines meet. Explain why the intersection point is an $\mathbb{F}$-point.

(ii) If $e_1 = 0$, show that the lines meet at an $\mathbb{F}$-point.
(iii) In (i), you needed to divide by $d_1e_2 - d_2e_1$ or something similar. Why can you do this?

3. [This illustrates case (b) of the proof of Lemma 6.3.5.]
Find where the $\mathbb{Q}$-circle

$$x^2 + y^2 - 4x - 2y - 4 = 0$$

and the $\mathbb{Q}$-line

$$x - y + 3 = 0$$

meet. For which α is this point of intersection a $\mathbb{Q}(\sqrt{\alpha})$-point?

4. Fill in the details of a general proof of case (b) of the proof of Lemma 6.3.5.

Additional Reading for Chapter 6

Many algebra textbooks contain a proof of the impossibility of the three geometrical constructions. You may like to compare our treatment with those in Bold (1969), Clark (1971), Fraleigh (1982), Gilbert (1976), Hadlock (1978) and Shapiro (1975). A self-contained account of Problem III is given in Hobson (1953).

The algebraic machinery developed in this book can also be used to characterize those regular polygons which can be constructed using straightedge and compass. See, for example, (Fraleigh, 1982, §48.2), (Clark, 1971, §135–138) and (Hadlock, 1978, §2.5 & 2.7).

Chapter 7
Zeros of Polynomials of Degrees 2, 3, and 4

In this chapter we begin our study of Problem IV, which is about solving polynomial equations, that is, finding the zeros of polynomials. We begin in familiar territory, namely solving quadratic equations, which you learned to do in high school. While thousands of years ago the Babylonians, Chinese, Egyptians, and Greeks knew how to find some solutions of some quadratic equations, it was not until the year 628 that an explicit, but not completely general, solution of $ax^2 + bx = c$ was produced by the Indian mathematician and astronomer Brahmagupta (590–668). Before continuing the discussion, it needs to be recognized that before the 16th century, there was a lack of understanding of not only imaginary numbers (and hence complex numbers) but even of negative integers. The complete solution of quadratic equations was published in 1594 by the Flemish mathematician Simon Stevin (1548–1620).

Cubic equations had also been examined by the Babylonians, Chinese, Egyptians, Greeks, and Indians and some solutions of particular cubic equations were known to them. However, it was not until the 16th century that the Italian mathematician Scipione del Ferro (1465–1526) discovered how to solve a wide class of cubic equations. As was the norm for that period, he did not publish his solution.

About 1540 the Italian mathematician (and later tax assessor) Lodovico Ferrari (1522–1565) discovered how to to solve quartic equations. The methods of solution for the general cubic and quartic equations were published for the first time in 1545 in the book "Artis Magnae" by the Italian mathematician, probabilist, and physician Gerolamo Cardano (1501–1576). (Artis Magnae was the tenth in a series of volumes, published by Cardano, called "Opus Perfectum" or "The Perfect Work".)

S. A. Morris et al., *Abstract Algebra and Famous Impossibilities*, Undergraduate Texts in Mathematics, https://doi.org/10.1007/978-3-031-05698-7_7

7.1 Solving Quadratic Equations

The Spanish Jewish mathematician Abraham bar Hiyya HaNasi (1070–1136) also known as Savasorda, published the first book in Europe to introduce Arabic algebra and, in particular, it included the almost complete solution of quadratic equations. It was written in Hebrew and in 1145 was translated into Latin as "Liber Embadorum". It became a widely used textbook in western European schools. In 1202 Leonardo of Pisa (1175–1250), known to us as Fibonacci, published "Liber Abaci" (Book of Calculation), Sigler (2003), which apparently contained all the knowledge possessed by the Arabs in algebra and arithmetic. In that book he popularized the Hindu-Arabic numeral system which had originated in India about the 6th century.

Let us consider the general quadratic equation

$$ax^2 + bx + c = 0, \text{ where } a, b, c \in \mathbb{R},\ a \neq 0. \tag{1}$$

Faced with the task of solving this, we might firstly say to ourselves that if the x term was not there, then we would have

$$ax^2 + c = 0 \implies x^2 = -c/a \implies x = \pm\sqrt{-c/a}.$$

Our approach, therefore, is to convert the quadratic equation to one without an x term.

So we shall solve (1) by firstly eliminating the x term.

Let y be defined by $x = y - \dfrac{b}{2a}$. Then (1) becomes

$$\begin{aligned}
&a\left(y - \frac{b}{2a}\right)^2 + by - \frac{b^2}{2a} + c = 0 \\
\implies &ay^2 + \frac{b^2}{4a} - \frac{b^2}{2a} + c = 0 \\
\implies &y^2 = \frac{b^2 - 4ac}{4a^2} \\
\implies &x = \frac{-b \pm \sqrt{b^2 - 4ac}}{2a}.
\end{aligned}$$

So we obtained the two solutions $x = \dfrac{-b \pm \sqrt{b^2 - 4ac}}{2a}$ of (1) by eliminating the x term in (1). This is called *depressing the equation*.

Exercises 7.1

1. Depress the quadratic equation $x^2 + 3x - 17 = 0$.
2. Solve the quadratic equation $x^2 + 5x + 4 = 0$ and verify that there are two roots, both of which are negative integers.
3. Solve the quadratic equation $4x^2 + 5x + 1 = 0$ and verify that there are two roots, both of which are negative rational numbers.
4. Solve the quadratic equation $4x^2 + 10x + 1 = 0$ and verify that there are two roots, one of which is a negative irrational number and the other is a positive irrational number.
5. Solve the quadratic equation $x^2 - 4x + 4 = 0$ and verify that there are two equal positive integer roots.
6. Solve the quadratic equation $3x^2 + 5x + 4 = 0$ and verify that there are two roots, both of which are non-real complex numbers.

7.2 Solving Cubic Equations

Having obtained the method for solving quadratic equations that you first met in high school, let us apply a similar technique to solve the general cubic equation

$$a_3x^3 + a_2x^2 + a_1x + a_0 = 0, \text{ where } a_1, a_2, a_3 \in \mathbb{R},\ a_3 \neq 0$$

Obviously there is no loss of generality in assuming $a_3 = 1$, since if it is not then we divide all terms by a_3. So we shall consider the cubic equation:

$$x^3 + a_2x^2 + a_1x + a_0 = 0. \tag{2}$$

Let us depress this cubic equation by eliminating the x^2 term.
Noting that $(x - b)^3 = x^3 - 3bx^2 + 3b^2x - b^3$, for any $b \in \mathbb{R}$, we can eliminate the x^2 in (2) by putting $x = y - \dfrac{a_2}{3}$ to obtain

$$\left(y - \frac{a_2}{3}\right)^3 + a_2\left(y - \frac{a_2}{3}\right)^2 + a_1\left(y - \frac{a_2}{3}\right) + a_0 =$$
$$\left[y^3 - a_2y^2 + \frac{{a_2}^2}{3}y - \frac{{a_2}^3}{27}\right] + \left[a_2y^2 - \frac{2a_2^2}{3}y + \frac{{a_2}^3}{9}\right] + \left[a_1y - \frac{a_1a_2}{3}\right] + a_0 = 0.$$

So gathering terms we have

$$y^3 + \left[a_1 - \frac{{a_2}^2}{3}\right]y + \left[a_0 + \frac{2a_2^3 - 9a_1a_2}{27}\right] = 0$$

So it has reduced to an equation of the form

$$y^3 + c_1 y + c_0 = 0. \tag{3}$$

where $c_1 = a_1 - \frac{a_2{}^2}{3}$ and $c_0 = a_0 + \frac{2a_2^3 - 9a_1 a_2}{27}$.

If $c_1 = 0$, that is $a_1 - \frac{a_2{}^2}{3} = 0$, then $y^3 = -c_0$, and so $y = -\sqrt[3]{c_0}$ and the cubic equation (3) is solved.

So we shall focus on the case where $c_1 \neq 0$.

Before we proceed with how to solve cubic equations, we look firstly at its history. But even before that we need to understand that at the beginning of the 16th century not only were complex numbers not known but even negative numbers were not known. Indeed, in Europe the number 0 was not known. While the symbol 0 appeared in early manuscripts, the first known use of it as a number is in the Bakhshali Manuscript held in the University of Oxford's Bodleian Library and the manuscript, written on birch bark, was recently carbon dated to the 3rd or 4th century. It is the oldest extant manuscript on Indian mathematics.

It was in the context of solving cubic equations that Gerolamo Cardano, in some sense, introduced imaginary numbers (and used negative numbers which he referred as fictitious numbers). However, it took a couple of hundred years before complex numbers were fully understood and accepted through the efforts of the Italian Rafael Bombelli (1526–1572), the Norwegian-Dane Caspar Wessel (1745–1818), and the Irishman William Rowan Hamilton (1805–1865). In 1702 Gottfried Wilhelm Leibniz (1646–1716) referred to complex numbers as follows: "wonder of analysis, a monstrosity of the human imagination".

Franciscan Friar Luca Bartolomeo de Pacioli (c.1447–1517) was a Florentine mathematician who wrote books in Italian, rather than Latin, and the books were illustrated by Leonardo da Vinci. Amongst these was "Summa de arithmetica, geometria, proportioni et proportionalita" (Summary of arithmetic, geometry, proportions and proportionality). In it he asserts that a solution to the cubic equation is as impossible as squaring the circle.

One of the earliest advances of mathematics in Europe after those of the Arabs and the Greeks, was that of the Italian mathematician Scipione del Ferro (1465–1526), who was a Lecturer in Arithmetic and Geometry at the University of Bologna. He discovered a method of solving cubic polynomial equations of a certain type, namely depressed cubic equations. As was generally the case at that time, del Ferro did not publish his results.

Niccoló Fontana of Brescia (1506–1557), known as Tartaglia, heard of del Ferro's result from a student of del Ferro, and in 1541 was able to come up with a method of proving del Ferro's result. Niccolò also kept his method secret.

In 1543, after Ferro's death, Gerolamo Cardano (1501–1576) and Lodovico Ferrari (1522–1565) (one of Cardano's students) travelled to Bologna to meet Hannival Nave, the son-in-law and successor at the University of Bologna of del Ferro, and they accessed his late father-in-law's notebook, where the solution to the depressed cubic equation appeared.

In the book Cardano (1993) published by Cardano in 1545 under the title Artis Magnae, Sive de Regulis Algebraicis Liber Unus (Book number one about The Great Art, or The Rules of Algebra) the following (translated from Latin) appeared: "Scipione Ferro of Bologna, almost thirty years ago, discovered the solution of the cube and things equal to a number [which in today's notation is the case $y^3 + c_1 y + c_0 = 0$], a really beautiful and admirable accomplishment. In distinction this discovery surpasses all mortal ingenuity, and all human subtlety. It is truly a gift from heaven, although at the same time a proof of the power of reason, and so illustrious that whoever attains it may believe himself capable of solving any problem. In emulation of him, my friend Nicolo Tartaglia wanting not to be outdone, solved the same case when he got into a contest with a pupil of Scipione, and moved by my many entreaties, gave it to me." Despite promising Tartaglia he would not publish the method, Cardano published the result in his book as Scipione had priority. Tartaglia was furious. (See Brooks (2017); Stewart (2017).)

In his book Cardano has the following approach to solve (3). Our presentation follows Bewersdorff (2006). Firstly observe that

$$(u+v)^3 = 3uv(u+v) + u^3 + v^3. \tag{4}$$

So comparing (3) and (4) and putting $y = u + v$, we see that

$$3uv = -c_1 \text{ and } u^3 + v^3 = -c_0.$$

$$\begin{aligned}
&\implies v = -\frac{c_1}{3u} \quad (\text{as } c_1 \neq 0 \implies u \neq 0 \text{ and } v \neq 0).\\
&\implies u^3 - \frac{c_1^3}{27u^3} = -c_0.\\
&\implies (u^3)^2 + c_0(u^3) - \frac{c_1^3}{27} = 0.
\end{aligned}$$

This is a quadratic equation in u^3, which we can solve as previously indicated:

$$u^3 = -\frac{c_0}{2} \pm \sqrt{\frac{c_0^2}{4} + \frac{c_1^3}{27}},$$

$$\implies u = \sqrt[3]{-\frac{c_0}{2} + \sqrt{\frac{c_0^2}{4} + \frac{c_1^3}{27}}} \text{ or } u = \sqrt[3]{-\frac{c_0}{2} - \sqrt{\frac{c_0^2}{4} + \frac{c_1^3}{27}}}.$$

But we know that $u^3 + v^3 = -c_0$ and so $v^3 = -u^3 - c_0$,

$$\implies v^3 = -\frac{c_0}{2} - \sqrt{\frac{c_0^2}{4} + \frac{c_1^3}{27}} \text{ or } v^3 = -\frac{c_0}{2} + \sqrt{\frac{c_0^2}{4} + \frac{c_1^3}{27}}$$

Thus

$$u = \sqrt[3]{-\frac{c_0}{2} + \sqrt{\frac{c_0^2}{4} + \frac{c_1^3}{27}}} \text{ and } v = \sqrt[3]{-\frac{c_0}{2} - \sqrt{\frac{c_0^2}{4} + \frac{c_1^3}{27}}} \tag{5}$$

or

$$u = \sqrt[3]{-\frac{c_0}{2} - \sqrt{\frac{c_0^2}{4} + \frac{c_1^3}{27}}} \text{ and } v = \sqrt[3]{-\frac{c_0}{2} + \sqrt{\frac{c_0^2}{4} + \frac{c_1^3}{27}}}. \tag{6}$$

As we know $y = u + v$, (5) and (6) show that

$$y = \sqrt[3]{-\frac{c_0}{2} - \sqrt{\frac{c_0^2}{4} + \frac{c_1^3}{27}}} + \sqrt[3]{-\frac{c_0}{2} + \sqrt{\frac{c_0^2}{4} + \frac{c_1^3}{27}}}. \tag{7}$$

So we have solved the cubic equation (3).

Care needs to be taken in interpreting the solution (7).
Each cube root has three possible values (some of which may be equal) but determining the particular cube root for u determines the particular cube root for v since we saw above that $3uv = -c_1$.

Finally, we know that $x = y - \frac{a_2}{3}$, $c_0 = a_0 + \frac{2a_2^3 - 9a_1a_2}{27}$, and $c_1 = a_1 - \frac{{a_2}^2}{3}$.
So we have also solved the general cubic equation (2).

Exercises 7.2

1. Show that the cubic equation $x^3 - 6x^2 + 11x - 6 = 0$ has three distinct roots, each of which is a positive integer.
2. Show that the cubic equation $x^3 - 7x^2 + 15x - 9 = 0$ has 2 equal roots, and a third root, all of which are positive integers.
3. Show that the cubic equation $x^3 + 2x^2 + x - 4 = 0$ has three unequal roots, one of which is a positive integer and the other two are non-real complex numbers.
4. Show that the cubic equation $5x^3 - 2x^2 + 5x - 2 = 0$ has one root which is a positive rational number and two unequal roots, each of which is an imaginary number.
5. Show that the cubic equation $3x^3 - 2x^2 - 6x + 4 = 0$ has three roots which are real numbers, one a rational number, and two which are unequal irrational numbers.

6. Show that the cubic equation $x^3 + 3x^2 - 3x + 5 = 0$ has three unequal roots only one of which is a real number and that real number is

$$-1 - \frac{2}{\sqrt[3]{5 - \sqrt{17}}} - \sqrt[3]{5 - \sqrt{17}}.$$

7. Depress the cubic equation $x^3 + 4x^2 + 5x + 1 = 0$.

7.3 Solving Quartic Equations

Now we turn to the problem of solving the general quartic equation:

$$a_4x^4 + a_3x^3 + a_2x^2 + a_1x + a_0 = 0\,, \text{ where } a_1, a_2, a_3, a_4 \in \mathbb{R} \text{ and } a_4 \neq 0.$$

As usual there is no loss of generality in assuming $a_4 = 1$, so we are looking to solve

$$x^4 + a_3x^3 + a_2x^2 + a_1x + a_0 = 0. \tag{8}$$

The solution was obtained in 1540 by the Italian Lodovico Ferrari, who was educated in mathematics by Gerolamo Cardano but the solution of the general quartic equation depended on the solution of the general cubic equation and so could not be published before Cardano's book appeared.

The first step in the proof is to depress the quartic equation (8); that is, we eliminate the x^3 term. We do this by putting $x = y - \dfrac{a_3}{4}$ to obtain an equation of the form:

$$y^4 + b_2y^2 + b_1y + b_0 = 0\,, \tag{9}$$

where b_2, b_1, and b_0 are known in terms of a_3, a_2, a_1, and a_0.

Once again our presentation follows Bewersdorff (2006). We add $2zy^2 + z^2$ to both sides of (9), where the value of z will be determined later, to obtain

$$y^4+2zy^2 + z^2=2zy^2 + z^2-[b_2y^2 + b_1y + b_0] = (2z - b_2)y^2 - b_1y + (z^2 - b_0). \tag{10}$$

We shall choose z so both sides of (10) become perfect squares. From (10) we have

$$(y^2 + z)^2 = \left(y\sqrt{2z - b_2} - \frac{b_1}{2\sqrt{2z - b_2}}\right)^2 + \left(z^2 - b_0 - \frac{b_1^2}{4(2z - b_2)}\right). \tag{11}$$

So we choose z such that

$$z^2 - b_0 - \frac{b_1^2}{4(2z - b_2)} = 0. \tag{12}$$

(12) gives us the cubic equation in (13) for z:

$$z^3 - \frac{b_2}{2}z^2 - b_0 z + \frac{b_2 b_0}{2} - \frac{b_1^2}{8} = 0 \tag{13}$$

We can now use the method described in Section 7.2 to solve the cubic equation (13) for z in terms of the known b_0, b_1, and b_2. We shall not bother writing this out in detail.

Having found z, we see from (11) and (12) that

$$y^2 = -z \pm \left(y\sqrt{2z - b_2} - \sqrt{z^2 - b_0}\right) \tag{14}$$

7.3.4 Remarks. Firstly, we note that it is a little surprising that the solution of the quartic equation involves the solution of a cubic equation and so gives cube roots in the solution.
Secondly, and more importantly, we observe that

> all solutions of the quadratic equations, the cubic equations, and the quartic equations involve only the operations of addition, subtraction, multiplication, division, square roots, cube roots, and fourth degree roots of the coefficients a_0, a_1, a_2, and a_3.

So it was very natural to ask if the solutions for all polynomial equations involve only addition, subtraction, multiplication, division and radicals of the given coefficients. This question was seriously investigated by prominent mathematicians for over 250 years before, as we shall see in Chapter 9, it was finally answered in the negative by the Norwegian mathematician Niels Henrik Abel (1802–1829) and the Italian mathematician Paolo Ruffini (1765–1822). For much of that period it had been assumed that the answer was in the positive but that the failure to find solutions for even the general quintic equation resulted from the complication increasing as the degree of the polynomial equation increased.

We shall discuss quintic equations and higher degree polynomial equations in Chapter 8 and Chapter 9.

Exercises 7.3

1. Depress the quartic equation $x^4 + 3x^3 + 2x^2 + 1 = 0$.
2. Show that the quartic equation $x^4 - 10x^3 + 35x^2 - 50x + 24 = 0$ has 4 distinct roots which are positive integers.
3. Show that the quartic equation $x^4 - 5x^3 + 7x^2 - 5x + 6 = 0$ has two roots which are positive integers and two distinct roots which are imaginary numbers.

4. Show that the quartic equation $x^4 + x^3 - 9x^2 + 11x - 10 = 0$ has two roots which are real numbers, one of which is $x = 2$, and two distinct roots which are non-real complex numbers.

Additional Reading for Chapter 7

To attack the problems of solving quadratic equations, cubic equations, and quartic equations we followed the presentation in Bewersdorff (2006) which is to depress the equation and then use "brute force". For the history of the solution of the cubic equation see Guilbeau (1930). As stated earlier, the method of solving cubic equations and quartic equations first appeared in print in Cardano (1993) which was published in Latin and there are translations in French, English and German. We also mentioned earlier that it was Cardano who explained and popularised negative numbers and complex numbers, without which a full understanding even of the general solution of quadratic equations is not possible. Also worth mentioning is that in Cardano's book the concept of a multiple root was first introduced. Solving quadratic, cubic, and quartic equations is discussed in Birkhoff and MacLane (1953).

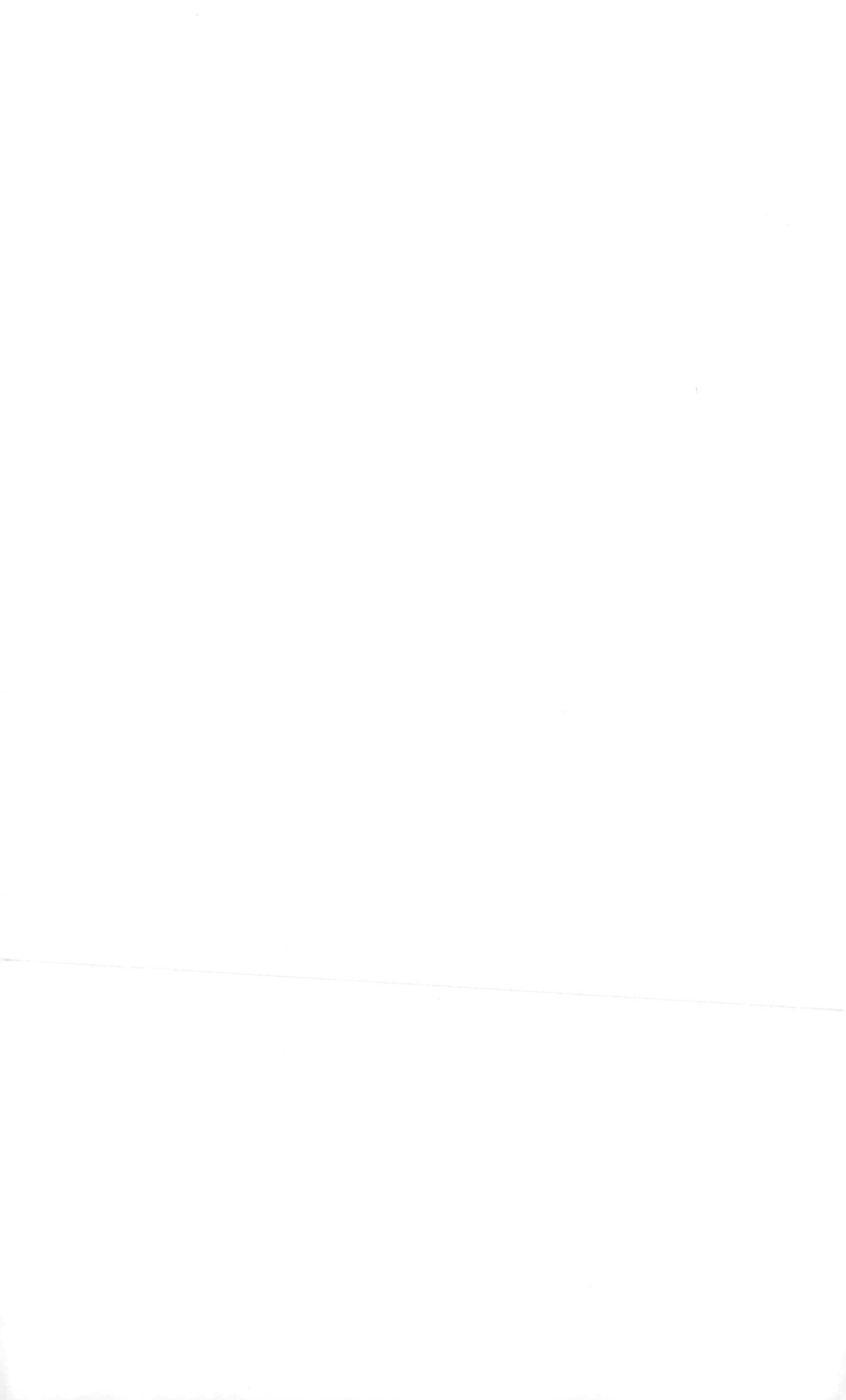

Chapter 8
Quintic Equations I: Symmetric Polynomials

In this chapter we begin the discussion of the Famous Impossibility which concerns finding solutions of a particular kind of the general quintic equation. The particular kind of solution is one which can be obtained from the coefficients of the polynomial equation of degree 5 (or more) using only addition, subtraction, multiplication, division, and radicals such as square roots, cube roots etc.

This chapter is devoted primarily to the study of symmetric polynomials, which are key to the proof of this impossibility.

Later we shall see that symmetric polynomials are also key to our proof that π is a transcendental number which completes the proof of the impossibility of Squaring the Circle.

8.1 Brief History of the Quintic Equation: 1683–1826

The first big step towards finding a method to solve quintic equations and equations of higher degree was made by the German mathematician Ehrenfried Walther von Tschirnhaus (1651–1708). In the paper, von Tschirnhaus (1683), called "A method for removing all intermediate terms from a given equation", he introduced what has become known as the Tschirnhaus transformation.

In Chapter 7 we described methods for solving quadratic equations, cubic equations, and quartic equations. We used a simple substitution for x to eliminate the term containing the second highest power of x and so obtain the depressed equation. Tschirnhaus showed that by using a somewhat more complicated substitution, one can eliminate the second highest power of x, the third highest power of x, the fourth highest power of x etc. These substitutions are called *Tschirnhaus transformations*.

S. A. Morris et al., *Abstract Algebra and Famous Impossibilities*, Undergraduate Texts in Mathematics, https://doi.org/10.1007/978-3-031-05698-7_8

To eliminate the second and third highest powers of x he put $x^2 = bx + y + a$; to eliminate the second, third, and fourth highest powers of x, he put $x^3 = cx^2 + bx + y + a$; and so on. He suggested that by such transformations one could solve polynomial equations of all degrees (presumably using only multiplication, division, addition, subtraction and radicals). Unfortunately if we try to apply the Tschirnhaus transformation for a general quintic, we end up en route having to solve a sextic (6th-degree) polynomial equation.

In 1786, the Swedish mathematician Erland Samuel Bring (1736–1798) found a way to avoid that by using a quartic substitution: $x^4 = y^4 + b_3y^3 + b_2y^2 + b_1y + b_0$. This extra parameter avoids increasing the degree en route. (See Harley (1864).)

The British mathematician George Birch Jerrard (1804–1863) apparently did not know of Bring's work, and developed an approach to do this for quintics and all higher degree polynomials. (See Jerrard (1859).) For a description of the Bring-Jerrard method see Adamchik and Jeffrey (2003). Jerrard claimed that his methods enabled one to solve quintic equations without in the process needing to solve any polynomial equation of degree greater than four.

At the meeting of the British Association for the Advancement of Science at Dublin in 1835 William Rowan Hamiltion in Hamilton (1836) presented an announcement by Jerrard of his work on polynomial equations, and commented briefly on it. Hamilton subsequently presented a more detailed report on Jerrard's Method at the next meeting of the British Association, at Bristol in 1836. Hamilton found that Jerrard had indeed constructed a general method for transforming polynomial equations to simpler forms by means of suitable Tschirnhaus transformations, but that the transformations only yielded a non-trivial result if the degree of the original polynomial equation was sufficiently large. Jerrard's methods proved to be of no assistance in solving the general quintic equation.

About 1770 Joseph-Louis Lagrange (1736–1813) related the various approaches that had been used to solve polynomial equations to the theory of groups of permutations. This groundbreaking work by Lagrange was a first step towards Galois theory, and the failure to develop methods of solution for equations of fifth and higher degrees suggested that such methods might not be possible.

By the end of the 18th century, the German mathematician Carl Friedrich Gauss (1777–1855), probably the greatest mathematician of that time, wrote in Gauss (1966, Latin version appeared in 1801): "there is little doubt that this problem does not so much defy modern methods of analysis as that it proposes the impossible". Early in the 19th century he wrote "After the labours of many geometers left little hope of ever arriving at the resolution of the general equation algebraically, it appears more and more likely that this resolution is impossible and contradictory.... Perhaps it will not be so difficult to prove, with all rigour, the impossibility for the fifth degree."

In 1798, the Italian mathematician Paolo Ruffini (1765–1822) claimed in his book "Teoria generale delle equazioni" to have proved that the general polynomial equation $x^n + a_{n-1}x^{n-1} + \ldots a_1x + a_0 = 0$, where $a_{n-1}, \ldots, a_0 \in \mathbb{Z}$, has no general solution in radicals if $n \geq 5$. His proof was difficult, and not generally accepted as correct. Subsequently it was found to have a gap (namely assuming that the roots were related to the coefficients of the polynomial in a particular way).

In 1826 the Norwegian mathematician Niels Henrik Abel (1802–1829), after whom abelian groups are named, gave a correct proof for the result which has become known as the Abel-Ruffini Theorem.

A proof of the Abel-Ruffini Theorem can be found in many textbooks. However, most of these use a method arising from Galois Theory which shows that a polynomial equation is solvable in radicals if and only if the Galois group is solvable. (In the British world, solvable groups are called soluble groups.)

Abel first published his result in 1824 and with more detail in 1826. In 1824 the French mathematician Évariste Galois (1811–1832) was only 13 years old! Obviously Galois Theory was not known and solvable groups were not known. Indeed the concept of a normal subgroup, used in defining solvable groups, had not been introduced.

The magnificent paper of Galois was submitted to the French Academy of Sciences in 1830 but was rejected. It was not published until 1846, 14 years after his death.

The approach presented in this book does not use Galois Theory, but rather is closer to the approach of Abel but also uses the work of the German mathematician Leopold Kronecker (1823–1891).

8.2 Fundamental Theorem of Algebra

Firstly, let us remind ourselves of the *Fundamental Theorem of Algebra* 8.2.1. There were many attempted proofs of the Fundamental Theorem of Algebra over two centuries, including by Leonhard Euler (1707–1783), Joseph-Louis Lagrange (1736–1813), Pierre-Simon Laplace (1749–1827), and Carl Friedrich Gauss (1777–1855), before the French amateur mathematician and bookstore manager Jean-Robert Argand (1768–1822) published the first rigorous proof in 1806. We do not include a proof of this well-known result here as all known proofs require some concepts outside algebra, such as continuity of functions. For proofs using complex analysis, see (Birkhoff and Maclane, 1953, §3 of Chapter V) or (Fraleigh, 1982, Theorem 31.18), and for a proof using topology, see (Morris, 2020, Theorem 8.5.1). We outline the topological proof in Morris (2020) in Exercises 8.2.

8.2.1 Theorem. [Fundamental Theorem of Algebra] *Every polynomial* $f(X) = a_n X^n + a_{n-1}X^{n-1} + \cdots + a_1 X + a_0$, *where each* a_i *is a complex number,* $a_n \neq 0$, *and* $n \geq 1$, *has a zero; that is, there exists a complex number* X_0 *such that* $f(X_0) = 0$.

If r is a zero of $f(X)$, then $f(X) = (X - r)g(X)$, where

$$g(X) = b_{n-1}X^{n-1} + b_{n-2}X^{n-2} + \cdots + b_1 X + b_0 \,,$$

where $b_{n-1}, b_{n-2}, \ldots, b_0 \in \mathbb{C}$, and $b_{n-1} \neq 0$. Using this we can easily apply mathematical induction to verify the following result.

8.2.2 Theorem. *Every polynomial*

$$f(X) = a_n X^n + a_{n-1} X^{n-1} + \cdots + a_1 X + a_0 ,$$

where each a_i is a complex number, $a_n \neq 0$, and $n \geq 1$, has n zeros in $\mathbb{C}$, some of which may be equal. Further,

$$f(X) = a_n (X - r_1)(X - r_2) \dots (X - r_n), \quad \textit{where } r_1, r_2, \dots, r_n \in \mathbb{C}.$$

So every quintic equation $a_5 X^5 + a_4 X^4 + a_3 X^3 + a^2 X^2 + a_1 X + a_0 = 0$, where $a_0, a_1, a_2, a_3, a_4, a_5 \in \mathbb{C}$ with $a_5 \neq 0$, has 5 roots, some of which may be equal.
When people say the general quintic equation does not have a solution, they mean that it does not have a solution *of a particular type*.
The particular type of solution they are referring to is one which can be expressed in terms of $a_0, a_1, a_2, a_3, a_4, a_5$ using only addition, subtraction, multiplication, division, and *radicals* such as $\sqrt{a_4}$, $\sqrt[3]{\frac{a_3}{a_2}}$, $\sqrt[5]{a_2 - a_1}$, etc.

Exercises 8.2

This exercise guides you through a topological proof of the Fundamental Theorem of Algebra 8.2.1

1.* Let $f(X) = a_n X^n + a_{n-1} X^{n-1} + \cdots + a_1 X + a_0$, where each a_i is a complex number, $a_n \neq 0$, and $n \geq 1$,

(i) Verify that for $R = |a_{n-1}| + |a_{n-2}| + \cdots + |a_0|$,
$|f(X)| \geq |X|^{n-1}$, for $|X| \geq \max\left\{1, \dfrac{R+1}{|a_n|}\right\}$.
[Hint. Use
$|f(X)| \geq |a_n||X|^n - |X|^{n-1}\left[|a_{n-1}| + \dfrac{|a_{n-2}|}{|X|} + \cdots + \dfrac{|a_0|}{|X^{n-1}|}\right]$.]

(ii) Verify that if we put $p_0 = |f(0)| = |a_0|$ then, by (i), there exists a $T > 0$ such that

$$|f(X)| > p_0, \quad \text{for all } |X| > T. \tag{1}$$

(iii) We shall **assume** two results from topology:

(a) the closed bounded subset $D = \{X : X \in \mathbb{C}, |X| \leq T\}$ is, by the Generalized Heine-Borel Theorem (Morris, 2020, Theorem 8.3.3), what is called a compact set (Morris, 2020, Definitions 7.1.7).
Thus by (Morris, 2020, Proposition 7.2.14):

(b) there exists an $X_0 \in D$ such that the continuous function $|f|$ from this compact set D to $\mathbb{R}$ has a least value at X_0; that is, $|f(X_0)| \leq |f(X)|$, for all $X \in D$.

Deduce from this and (i) that $|f(0)| \geq |f(X_0|$ and so

$$|f(X_0)| \leq |f(X)|, \quad \text{for all } X \in \mathbb{C}. \tag{2}$$

(iv) To prove the Fundamental Theorem of Algebra 8.2.1, we are required to show that $f(X_0) = 0$. Putting $P(X) = f(X + X_0)$, our problem now becomes that of proving that $P(0) = 0$.
Verify that (2) implies that

$$|P(0)| \leq |P(X)|, \text{ for all } X \in \mathbb{C}. \tag{3}$$

(v) Put $P(X) = b_n X^n + b_{n-1}X^{n-1} + \cdots + b_0$, where $b_0, b_1, \ldots, b_n \in \mathbb{C}$. So $P(0) = b_0$. Our task, then, is to show that $b_0 = 0$.
Suppose $b_0 \neq 0$. Verify that

$$P(X) = b_0 + b_k X^k + X^{k+1}Q(X), \tag{4}$$

where $Q(X)$ is a polynomial and k is the smallest $i > 0$ with $b_i \neq 0$.

(vi) As an example of (v), find $Q(X)$ if

$$P(X) = 10X^7 + 6X^5 + 3X^4 + 4X^3 + 2X^2 + 1.$$

(vii) Let $w \in \mathbb{C}$ be a kth root of the number $-b_0/b_k$; that is, $w^k = -b_0/b_k$.
Noting that $Q(X)$ is a polynomial, verify that for t a real number,

$$t\,|Q(tw)| \to 0, \text{ (that is, } t|Q(tw)| \text{ converges to 0)} \quad \text{as } t \to 0$$

and that this implies that $t\,|w^{k+1}Q(tw)| \to 0 \quad \text{as } t \to 0$.

(viii) Deduce from (vii) that there exists a real number t_0 with $0 < t_0 < 1$ such that

$$t_0\,|w^{k+1}Q(t_0 w)| < |b_0| \tag{5}$$

(ix) Verify from (4) that

$$P(t_0 w) = b_0(1 - t_0{}^k) + (t_0 w)^{k+1}Q(t_0 w). \tag{6}$$

(x) Using (ix), (5) and the definition of w, prove that

$$\begin{aligned} |P(t_0 w)| &\leq (1 - t_0{}^k)|b_0| + t_0{}^{k+1}|w^{k+1}Q(t_0 w)| \\ &< (1 - t_0{}^k)\,|b_0| + t_0{}^k\,|b_0|, \quad \text{by (6)} \\ &= |b_0| \\ &= |P(0)| \end{aligned} \tag{7}$$

(xi) Observe that (7) contradicts (3).
Therefore our supposition that $b_0 \neq 0$ is false. Hence $P(0) = 0$, which completes the proof of the Fundamental Theorem of Algebra 8.2.1.

8.3 Primitive and Symmetric Polynomials

Our overall aim is to prove that some quintic equations with integer coefficients do not have a solution involving only addition, subtraction, multiplication, division and radicals of the coefficients of the quintic equation. So we shall focus our attention on polynomials with integer or even rational number coefficients.

8.3.1 Proposition. *Let $p(X) = a_n X^n + a_{n-1} X^{n-1} + \cdots + a_1 X + a_0$ be a polynomial with integer coefficients and $a_n \neq 0$. If the rational number $\frac{r}{s}$ is a root of the polynomial equation $p(X) = 0$, where $\frac{r}{s}$ is in its lowest terms (i.e., there are no nontrivial common factors of r and s), then $r | a_0$ and $s | a_n$, where $|$ denotes divides.*

Proof. As $\frac{r}{s}$ is a root, $p\left(\frac{r}{s}\right) = 0$, that is

$$s^n p\left(\frac{r}{s}\right) = 0 = a_n r^n + a_{n-1} r^{n-1} s + \cdots + a_1 r s^{n-1} + a_0 s^n.$$

$$\text{So } -a_n r^n = s(a_{n-1} r^{n-1} + a_{n-2} r^{n-2} s + \cdots + a_1 r^{n-1} s^{n-2} + a_0 s^{n-1}).$$

Thus $s | a_n r^n$. As r and s have no common factors, r^2 and s have no common factors which by mathematical induction yields that r^n and s have no common factors. Hence $s | a_n$.

$$\text{Similarly } -a_0 s^n = r(a_n r^{n-1} + a_{n-1} r^{n-2} s + \cdots + a_1 s^{n-1}).$$

So $r | a_0 s^n$ which implies, by an analogous argument, that $r | a_0$. ■

8.3.2 Corollary. [Gauss Lemma 1] *Any rational number zero of a monic polynomial with integer coefficients (that is, a polynomial in $\mathbb{Z}[X]$) is an integer.* ■

8.3.3 Definition. The polynomial $p(X) = a_n X^n + a_{n-1} X^{n-1} + \cdots + a_0 \in \mathbb{Z}[X]$ is said to be *primitive* if its coefficients $a_0, a_1, \ldots, a_n$ have greatest common divisor equal to 1. ■

8.3.4 Examples. The polynomial $2 - 3X^2$ is primitive, while the polynomial $2 - 4X^2$ is not primitive. ■

We now show that the product of two primitive polynomials is a primitive polynomial.

8.3.5 Proposition. *If* $f(X) = a_nX^n + a_{n-1}X^{n-1} + \cdots + a_1X + a_0$ *and* $g(X) = b_mX^m + b_{m-1}X^{m-1} + \cdots + b_1X + b_0$ *are primitive polynomials, then* $p(X) = f(X)g(X) = c_{n+m}X^{n+m} + c_{n+m-1}X^{n+m-1} + \cdots + c_1X + c_0$ *is a primitive polynomial.*

Proof. Suppose there exists a prime number p which divides $c_0, c_1, \ldots, c_{n+m-1}$ and c_{n+m}. Consider $a_0, a_1, \ldots, a_n$ and let a_i be the first in this list which is not divisible by p. The coefficient a_i certainly exists as not all $a_0, a_1, \ldots, a_n$ are divisible by p, because $f(X)$ is primitive. Similarly let b_j be the first of $b_0, b_1, \ldots, b_m$ which is not divisible by p. Now

$$a_ib_j = c_{i+j} - [a_0b_{i+j} + \cdots + a_{i-1}b_{j+1} + a_{i+1}b_{j-1} + \cdots + a_{i+j}b_0]$$

from which it follows that a_ib_j is divisible by p since each term on the right hand side is divisible by p. So p must divide a_i or b_j, which is a contradiction. So our supposition is false. Therefore $c_0, c_1, \ldots, c_{n+m-1}, c_{n+m}$ have greatest common divisor 1. Thus $p(x)$ is a primitive polynomial. ■

8.3.6 Lemma. *If* $f(X)$ *is a non-zero polynomial in* $\mathbb{Q}[X]$*, then there exists a unique primitive polynomial* $f^*(X) \in \mathbb{Z}[X]$ *and a unique* $k_f \in \mathbb{Q}, k_f > 0$*, such that* $f(X) = k_f f^*(X)$*. Further, the degree of* $f(X)$ *equals the degree of* $f^*(X)$*.*

Proof. Let us write

$$f(X) = \frac{a_n}{b_n}X^n + \frac{a_{n-1}}{b_{n-1}}X^{n-1} + \cdots + \frac{a_0}{b_0},$$

for $b_n, b_{n-1}, \ldots, b_0, a_n, a_{n-1}, \ldots, a_0 \in \mathbb{Z}$. Of course $b_i \neq 0$, for $i = 0, 1, 2, \ldots, n$. Put $k = \dfrac{1}{b_0b_1 \ldots b_n}$, so that $k \in \mathbb{Q}$. Then we have

$$f(X) = kg(X), \text{ for } g(X) \in \mathbb{Z}[X].$$

Let k' be the greatest common divisor of the integral coefficients of $g(X)$. So $f^*(X) = \dfrac{g(X)}{k'}$ is a primitive polynomial in $\mathbb{Z}[X]$ and $f(X) = k_f f^*(X)$, where $k_f = kk'$.

To prove the uniqueness of $f^*(X)$ and k_f, it suffices to prove the uniqueness of $f^*(X)$.

If $f(X) = k_1f_1(X) = k_2f_2(X), k_1, k_2 > 0$, then $f_1(X) = \dfrac{k_2}{k_1}f_2(X)$. So $\dfrac{k_2}{k_1} = \dfrac{u}{v}$, where u, v are relatively prime positive integers. So $vf_1(X) = uf_2(X)$. Thus the coefficients of $uf_2(X)$ have v as a common factor. As u, v are relatively prime, this implies that v is a common divisor of all the coefficients of $f_2(X)$. But $f_2(X)$ is a primitive polynomial and so $v = 1$. Similarly $u = 1$.

It is clear from the construction of $f^*(X)$, that $f(X)$ and $f^*(X)$ have the same degree, which completes the proof. ■

Our next proposition generalizes Corollary 8.3.2 [Gauss Lemma 1].

8.3.7 Proposition. *Let $f(X) \in \mathbb{Z}[X]$. If $f(X)$ can be factored into polynomials with rational coefficients, then it can be factored into polynomials of the same degrees with integral coefficients.*

Proof. Let $f(X) = g(X)h(X)$, where $g(X), h(X) \in \mathbb{Q}[X]$. By Lemma 8.3.6, there exist positive integers c_g and c_h, and $g^*(X), h^*(X) \in \mathbb{Z}[X]$ such that

$$f(X) = g(X)h(X) = (c_g g^*(X))(c_h h^*(X)) = (c_g c_h)(g^*(X)h^*(X)),$$

where $g(X)$ and $g^*(X)$ have the same degree as $h(X)$ and $h^*(X)$, respectively.

As $f(X), g^*(X), h^*(X) \in \mathbb{Z}[X]$, and $f(X) = c_g c_h g^*(X)h^*(X)$ and from Lemma 8.3.6 $c_g, c_h > 0$, we have that $c_g c_h$ is a positive integer.

Then $f(X) = (c_g c_h g^*(X))h^*(X)$, where $c_g c_h g^*(X) \in \mathbb{Z}[X]$ and has the same degree as $g(X)$ and $h^*(X) \in \mathbb{Z}[X]$ and has the same degree as $h(X)$, which completes the proof. ∎

As an immediate Corollary we obtain:

8.3.8 Proposition. [Gauss Lemma 2] *Let $p(X) \in \mathbb{Z}[X]$. If $p(X)$ is irreducible over $\mathbb{Z}$, then it is irreducible over $\mathbb{Q}$.* ∎

The next proposition from Abel (1826), is somewhat surprising, and very useful.

8.3.9 Proposition. [Abel's Irreducibility Theorem] *Let $\mathbb{F}$ be any subfield of $\mathbb{C}$ and let $f(X) \in \mathbb{F}[X]$ be irreducible over $\mathbb{F}$. If $a \in \mathbb{C}$ is a zero of $f(X)$ and also a zero of $G(X) \in \mathbb{F}[X]$, then all the zeros of $f(X)$ are also zeros of $G(X)$. Further, $G(X) = f(X)G_1(X)$, where $G_1(X) \in \mathbb{F}[X]$.*

Proof. Let $h(X) \in \mathbb{F}[X]$ be the greatest common divisor of the two polynomials $f(X)$ and $G(X)$, so that, in particular, $f(X) = f^*(X)h(X)$ and $G(X)=G^*(X)h(X)$. Observe that $h(X) \neq 1$ as $(X - a)$ divides both $f(X)$ and $G(X)$. As $f(X)$ is irreducible over $\mathbb{F}$, this implies $f^*(X) = 1$ and so $f(X) = h(X)$. Thus $f(X)$ divides $G(X)$ and so all zeros of $f(X)$ are zeros of $G(X)$. ∎

8.3.10 Corollary. *Let a be a zero of the irreducible polynomial $f(X) \in \mathbb{F}[X]$ and also a zero of $G(X) \in \mathbb{F}[X]$, where $\mathbb{F}$ is a subfield of $\mathbb{C}$. If the degree of $G(X)$ is strictly less than that of $f(X)$, then all the coefficients of $G(X)$ are zero.* ∎

8.3.11 Corollary. *If $\mathbb{F}$ is a subfield of $\mathbb{C}$ and $f(X) \in \mathbb{F}[X]$ is irreducible over $\mathbb{F}$, then there is no other irreducible polynomial over $\mathbb{F}$ that has a common zero with $f(X)$.* ∎

8.3.12 Corollary. *Let $\mathbb{F}$ be a subfield of $\mathbb{C}$, $f(X) \in \mathbb{F}[X]$ an irreducible polynomial over $\mathbb{F}$ of degree n, for some positive integer n, and let $\alpha \in \mathbb{C}$ be a zero of $f(X)$. Then each member β of the extension field $\mathbb{F}(\alpha)$ can be represented as a polynomial of degree $(n-1)$ in α with coefficients in $\mathbb{F}$. Further, this representation is unique.*

Proof. We recall that $\mathbb{F}(\alpha)$ is a vector space over $\mathbb{F}$. So if $\beta \in \mathbb{F}(\alpha)$, $\beta = a_m\alpha^m + a_{m-1}\alpha^{m-1} + \cdots + a_0$, where each $a_i \in \mathbb{F}$.

But as α is a zero of $f(X)$ which is a polynomial of degree n, we can reduce all the powers of α in this representation of β to strictly less than n; that is,

$$\beta = b_{n-1}\alpha^{n-1} + b_{n-2}\alpha^{n-2} + \cdots + b_0, \text{ where each } b_i \in \mathbb{F}. \tag{8}$$

Further, the representation in (8) is unique, since if

$$b_{n-1}\alpha^{n-1} + b_{n-2}\alpha^{n-2} + \cdots + b_0 = c_{n-1}\alpha^{n-1} + c_{n-2}\alpha^{n-2} + \cdots + c_0\,,$$

then for $d_i = b_i - c_i$, $i = \{1, 2, \ldots, n-1\}$,

$$0 = d_{n-1}\alpha^{n-1} + d_{n-2}\alpha^{n-2} + \cdots + d_0$$

which says that the zero α of $f(X)$ is a zero of a polynomial of degree $n-1$ (or less), and so by Corollary 8.3.10, every $d_i = 0$. Thus each $b_i = c_i$, and so the representation is unique. ■

We now proceed to obtain a result known as the Fundamental Theorem on Symmetric Polynomials. This will allow us to deduce results such as: if $f(X)$ is a polynomial whose coefficients are all rational numbers, then every symmetric expression in the zeros of $f(X)$ – such as the sum of the squares of the zeros of $f(X)$ – is a rational number. There are various proofs of the Fundamental Theorem on Symmetric Polynomials. We present here a standard proof. (For the history and proofs of this result, see Blum-Smith and Coskey (2016); Edwards (1984).)

According to the British mathematician Derek Thomas Whiteside (1932–2008) who examined and published Newton's Notes, the result was known to Isaac Newton (1642–1726), but he published only part of it in 1707 in Newton (1769). It was generally known as Newton's Theorem, but a careful statement and proof did not appear in print until the 19th century. It was a key step in what became Galois Theory.

Firstly we shall introduce the elementary symmetric polynomials.

8.3.13 Definitions. The *elementary symmetric polynomials* in n indeterminates $X_1, X_2, \ldots, X_n$ written e_k, $k \in \{0, 1, 2, \ldots, n\}$, are defined by

$$e_0(X_1, \ldots, X_n) = 1$$

$$e_1(X_1, \ldots, X_n) = \sum_{1 \le j \le n} X_j$$

$$e_2(X_1, \ldots, X_n) = \sum_{1 \le j < k \le n} X_j X_k$$

$$e_3(x_1, \ldots, X_n) = \sum_{1 \le j < k < l \le n} X_j X_k X_l$$

$$\vdots$$

$$e_n(X_1, \ldots, X_n) = X_1 X_2 \ldots X_n.$$

If S is any set, then a *permutation* of S is a function $g : S \to S$ which is one-to-one and onto.
A polynomial $f(X_1, X_2, \ldots, X_n)$ in n indeterminates is said to be a *symmetric polynomial* if for every permutation $g : \{1, 2, \ldots, n\} \to \{1, 2, \ldots, n\}$,

$$f(X_1, X_2, \ldots, X_n) = f(X_{g(1)}, X_{g(2)}, \ldots, X_{g(n)}). \qquad \blacksquare$$

The next lemma contains two routine consequences of this definition. We leave the proof as an exercise.

8.3.14 Lemma. *(i) If $f(X_1, \ldots, X_n)$ and $g(X_1, \ldots, X_n)$ are symmetric polynomials in $X_1, \ldots, X_n$ then so are*

$$f(X_1, \ldots, X_n) + g(X_1, \ldots, X_n),$$
$$f(X_1, \ldots, X_n) - g(X_1, \ldots, X_n), \text{ and}$$
$$f(X_1, \ldots, X_n) g(X_1, \ldots, X_n),$$

(their sum, difference, and product).
(ii) If $h(Y_1, \ldots, Y_m)$ is **any** *polynomial in indeterminates $Y_1, \ldots, Y_m$ and if $g_1(X_1, \ldots, X_n), \ldots, g_m(X_1, \ldots, X_n)$ are symmetric polynomials in $X_1, \ldots, X_n$ then*

$$h\left(g_1(X_1, \ldots, X_n), \ldots, g_m(X_1, \ldots, X_n)\right)$$

is also symmetric in $X_1, \ldots, X_n$. $\blacksquare$

Note that "quotient" is not included in (i) of Lemma 8.3.14 since the quotient of two polynomials is in general not a polynomial.

8.3.15 Remark. *The elementary symmetric polynomials provide a connection between the zeros of a polynomial and its coefficients.*

To see this connection, let

$$f(Y) = a_n Y^n + a_{n-1} Y^{n-1} + \ldots + a_1 Y + a_0 \qquad (**)$$

be a polynomial of degree n (that is, $a_n \neq 0$) with indeterminate Y and coefficients in a field $\mathbb{F}$. Assume also that f has n zeros $\gamma_1, \gamma_2, \dots, \gamma_n$ in some (possibly larger) field $\mathbb{E}$ so that

$$f(Y) = a_n(Y - \gamma_1)(Y - \gamma_2)\dots(Y - \gamma_n).$$

It follows immediately from the definition of the elementary symmetric polynomial that this expands to

$$f(Y) = a_n Y^n - a_n e_1(\gamma_1, \dots, \gamma_n)Y^{n-1} + a_n e_2(\gamma_1, \dots, \gamma_n)Y^{n-2} - \dots + (-1)^n a_n e_n(\gamma_1, \dots, \gamma_n).$$

Thus from (**)

$$e_1(\gamma_1, \dots, \gamma_n) = -a_{n-1}/a_n,$$
$$e_2(\gamma_1, \dots, \gamma_n) = a_{n-2}/a_n,$$

and so on down to

$$e_n(\gamma_1, \dots, \gamma_n) = (-1)^n a_0/a_n.$$

Thus *each elementary symmetric polynomial evaluated at the zeros of (**) is plus or minus the quotient of a coefficient divided by the leading coefficient.* Note also that all of these quotients are in $\mathbb{F}$ (even though the γ's need not be in $\mathbb{F}$). ■

8.3.16 Remarks. Observe that if $f(X)$ is a monic polynomial of degree n with zeros $\alpha_1, \alpha_2, \dots, \alpha_n \in \mathbb{C}$, then $f(X)$ equals $\prod_{i=1}^{n}(X - \alpha_i)$ which in turn equals

$$X^n - e_1(\alpha_1, \dots, \alpha_n)X^{n-1} + e_2(\alpha_1, \dots, \alpha_n)X^{n-2} + \dots + (-1)^n e_n(\alpha_1, \dots, \alpha_n).$$

This is known as *Vieta's Theorem*, named after the French mathematician François Viéta (1540–1603).

We shall see that any symmetric polynomial $f(X_1, X_2, \dots, X_n)$ can be obtained from the elementary symmetric polynomials

$$e_0(X_1, X_2, \dots, X_n), e_1(X_1, X_2, \dots, X_n), \dots, e_n(X_1, X_2, \dots, X_n)$$

using only multiplication and addition. For example

$$f(X_1, X_2) = X_1^3 + X_2^3 - 5 = e_1(X_1, X_2)^3 - 3e_2(X_1, X_2)e_1(X_1, X_2) - 5. \quad ■$$

8.3.17 Theorem. [Fundamental Theorem on Symmetric Polynomials] *Every symmetric polynomial g, with coefficients in a field $\mathbb{F}$, in the indeterminates $X_1, X_2, \dots, X_n$ can be written as a polynomial h (with coefficients in $\mathbb{F}$) in the elementary symmetric polynomials. Moreover,*

(i) the degree of h is no more than the degree of g, and
(ii) if g has integer coefficients then so does h.

Before proving the theorem, we describe an algorithm for constructing such a polynomial h. Later we shall prove that the algorithm works; this will prove the theorem. We firstly need another definition.

8.3.18 Definition. Consider two nonzero terms

$$cX_1^{i_1}X_2^{i_2}\ldots X_n^{i_n} \quad \text{and} \quad dX_1^{j_1}X_2^{j_n}\ldots X_n^{j_n} \quad (c, d \in \mathbb{F}).$$

We say that the first of these is of *higher order than* the second (or, equivalently that the second is of *lower order than* the first) if, at the first position, say k, in which the i's and j's differ, we have $i_k > j_k$. ∎

For example, $X_1^2X_2^3X_3$ is of higher order than each of $X_1^2X_2X_3$ and $X_1X_2^5X_3^{10}$, while $X_1X_2^5X_3^{10}$ is of lower order than $X_1^2X_2^3X_3$.

8.3.19 Algorithm. [Fundamental Algorithm for Symmetric Polynomials] The aim of the algorithm is as follows: *Given a nonzero symmetric polynomial g in n indeterminates $X_1, X_2, \ldots, X_n$ satisfying the hypotheses of the Fundamental Theorem on Symmetric Polynomials, to construct a polynomial h satisfying the conclusion of that theorem.*

Step 0. Initially set $\ell = 1$ and $g_1 = g$ (which is not zero).

Step 1.

(a) Find the term of highest order in g_ℓ, say

$$cX_1^{i_1}X_2^{i_2}\ldots X_n^{i_n} \quad (c \in \mathbb{F}, c \neq 0).$$

(b) With c and $i_1, i_2, \ldots, i_n$ as in (a), set

$$h_\ell(Y_1, Y_2, \ldots, Y_n) = cY_1^{i_1-i_2}Y_2^{i_2-i_3}\ldots Y_{n-1}^{i_{n-1}-i_n}Y_n^{i_n}.$$

(c) Put

$$g_{\ell+1}(X_1, X_2, \ldots, X_n) = g_\ell(X_1, X_2, \ldots, X_n) - h_\ell(e_1, e_2, \ldots, e_n).$$

(d) Increase ℓ by 1.

(e) If $g_\ell \neq 0$ then go back to (a) and repeat Step 1; otherwise proceed to Step 2 below.

Step 2. Put $h = h_1 + h_2 + \ldots + h_{\ell-1}$. ∎

8.3.20 Example. We apply the above algorithm to the symmetric polynomial (in $n = 2$ indeterminates)

$$g(X_1, X_2) = 3X_1^3X_2 - 2X_1^2X_2^2 + 3X_1X_2^3.$$

(See Definitions 8.3.13 for formulae for e_1 and e_2.)
Step 0. Set $\ell = 1$ and put $g_1 = g$.
Step 1.

(a) The term of highest order in g_1 is $3X_1^3X_2$.
(b) $h_1(Y_1, Y_2) = 3Y_1^{3-1}Y_2^1 = 3Y_1^2Y_2$.
(c) $g_2(X_1, X_2) = g_1(X_1, X_2) - h_1(e_1, e_2)$
$= 3X_1^3X_2 - 2X_1^2X_2^2 + 3X_1X_2^3 - 3(X_1 + X_2)^2(X_1X_2)$
$= -8X_1^2X_2^2$.
(d) Set $\ell = 2$.
(e) As $g_2 \neq 0$, return to (a) and repeat Step 1 with g_2 in place of g_1.

Step 1.

(a) The term of highest order in g_2 is $-8X_1^2X_2^2$.
(b) $h_2(Y_1, Y_2) = -8Y_1^{2-2}Y_2^2 = -8Y_2^2$.
(c) $g_3(X_1, X_2) = g_2(X_1, X_2) - h_2(e_1, e_2) = -8X_1^2X_2^2 + 8X_1^2X_2^2 = 0$.
(d) Set $\ell = 3$.
(e) Since $g_3 = 0$ go on to Step 2.

Step 2. Set $h = h_1 + h_2$; that is, $h(Y_1, Y_2) = 3Y_1^2Y_2 - 8Y_2^2$.
As a check, note that

$$\begin{aligned} h(e_1, e_2) &= 3(X_1 + X_2)^2(X_1X_2) - 8(X_1X_2)^2 \\ &= 3X_1^3X_2 - 2X_1^2X_2^2 + 3X_1X_2^3 \\ &= g(X_1, X_2), \end{aligned}$$

and also that $\deg h = 3 < 4 = \deg g$ and that the polynomial h has integer coefficients. Thus h satisfies the conclusion of Theorem 8.3.17. ■

To prove that the algorithm works in every case we need the following lemmas.

8.3.21 Lemma. *Let $g(X_1, \ldots, X_n)$ be a symmetric polynomial over a field $\mathbb{F}$ and let*

$$cX_1^{i_1}X_2^{i_2}\ldots X_n^{i_n} \quad (c \in \mathbb{F},\ c \neq 0) \tag{9}$$

be the term of highest order in g. Then

$$i_1 \geq i_2 \geq \ldots \geq i_n. \tag{10}$$

Proof. Suppose $i_k < i_{k+1}$, contrary to the conclusion of the lemma. Then we apply to g the permutation ρ which just interchanges the indeterminates X_k and X_{k+1}. The resulting polynomial then contains the term

$$cX_1^{i_1} \dots X_{k-1}^{i_{k-1}} X_k^{i_{k+1}} X_{k+1}^{i_k} X_{k+2}^{i_{k+2}} \dots X_n^{i_n} \tag{11}$$

which is of higher order than (9). But g is a symmetric polynomial. so the term (11) must be in g, which contradicts our choice of (9) as the term of highest order in g. Thus our supposition is false and (10) holds. ∎

8.3.22 Lemma. *The term of highest order in the expansion of the polynomial* ${e_1}^{i_1-i_2}{e_2}^{i_2-i_3}\dots{e_{n-1}}^{i_{n-1}-i_n}{e_n}^{i_n}$, *where* $e_1, e_2, \dots, e_n$ *are the elementary symmetric polynomials, is* $X_1^{i_1} X_2^{i_2} \dots X_n^{i_n}$.

Proof. The term of highest order in a product of polynomials is the product of the terms of highest order in each polynomial. The terms of highest order in $e_1, e_2, \dots, e_n$ are respectively

$$X_1,\ X_1X_2,\ X_1X_2X_3, \dots, X_1X_2X_3\dots X_n.$$

Hence the term of highest order in the product is

$$X_1^{i_1-i_2}(X_1X_2)^{i_2-i_3}\dots(X_1X_2\dots X_n)^{i_n} = X_1^{i_1}X_2^{i_2}\dots X_n^{i_n}.$$ ∎

Proof of Theorem 8.3.17. To prove the theorem we prove that the Algorithm 8.3.19 always produces a polynomial h with the properties stated in the theorem.

If $g \neq 0$ is a symmetric polynomial in $X_1, X_2, \dots, X_n$, we let $D \in \mathbb{N}$ denote the number of monomials (not necessarily occurring in g) which are of lower order than the highest term occurring in g. (Recall that a *monomial* is a term $X_1^{i_1}\dots X_n^{i_n}$ with coefficient 1.) Our proof is by mathematical induction on D.

If $D = 0$ then g is a constant polynomial and it is left as an easy exercise to prove that the algorithm works in this case.

Now let $k \in \mathbb{N}$ and assume that the algorithm works for all symmetric polynomials g with $D \leq k$.

Consider now a polynomial g_1 with $D = k + 1$, and let its term of highest order be

$$m_1 = cX_1^{i_1}X_2^{i_2}\dots X_n^{i_n}, \tag{12}$$

where $c \in \mathbb{F}$ is nonzero.

We wish to prove that Algorithm 8.3.19 works for g_1.

Firstly we shall consider what happens with the first pass through Step 1 of the algorithm.

Then we shall consider the two cases $g_2 = 0$ in that algorithm and $g_2 \neq 0$ and proceed as instructed in the algorithm.

The first pass through Step 1 of the algorithm gives

$$h_1(Y_1, Y_2, \dots, Y_n) = cY_1^{i_1-i_2}Y_2^{i_2-i_3}\dots Y_n^{i_n} \tag{13}$$

and

$$g_2(X_1, X_2, \ldots, X_n) = g_1(X_1, X_2, \ldots, X_n) - h_1(e_1, e_2, \ldots, e_n). \qquad (14)$$

Now (13) defines h_1 as a polynomial since $i_1 \geq i_2 \geq \ldots \geq i_n$ by Lemma 8.3.21. Thus, by (ii) of Lemma 8.3.14, $h_1(\sigma_1, \ldots, \sigma_n)$ is symmetric in $X_1, \ldots, X_n$ and hence, by (i) of Lemma 8.3.14, so is g_2.

By Lemma 8.3.22, moreover, the two polynomials on the right hand side of (14) have the same term of highest order and so these terms cancel each other out in (14). If g_2 is zero, it is an easy exercise to check that the choice $h = h_1$ satisfies the requirements of the theorem. Hence we consider below only the case where $g_2 \neq 0$. If m_2 is the term of highest order in g_2, then m_2 is of lower order than m_1 (since the terms of highest order cancel each other out in (14)) and so the D for g_2 (the number of monomials of lower order than m_2) is less than $k + 1$ which is the D for g_1 (the number of monomials of lower order than m_1). Thus the inductive assumption applies to g_2 and so the algorithm gives a polynomial h^*, say, such that

$$g_2(X_1, X_2, \ldots, X_n) = h^*(e_1, e_2, \ldots, e_n).$$

Hence by (14), $g_1(X_1, X_2, \ldots, X_n) = h(e_1, e_2, \ldots, e_n)$ where $h = h_1 + h^*$.

To see that the "moreover" parts of the theorem are true, note simply that

(i) the degree of $h_1(Y_1, \ldots, Y_n)$ in (13) is i_1 while the degree of g_1 is at least $i_1 + i_2 + \ldots + i_n$, so that the degree of h_1 is not more than the degree of g_1, and
(ii) if g_1 has integer coefficients then c in (12) is an integer, which means, by (13), that h_1 has integer coefficients.

The "moreover" results now follow easily. (Formally, each of these results requires a separate proof by induction.) ■

8.3.23 Corollary. *Let $\mathbb{F}$ be a field and let $f(X)$ be a polynomial of degree n with coefficients in $\mathbb{F}$ and with n zeros $\alpha_1, \alpha_2, \ldots, \alpha_n$ in some extension field $\mathbb{E}$ of $\mathbb{F}$. If g is any symmetric polynomial in $X_1, X_2, \ldots, X_n$ with coefficients in $\mathbb{F}$ then $g(\alpha_1, \alpha_2, \ldots, \alpha_n) \in \mathbb{F}$.*

Proof. By the Fundamental Theorem on Symmetric Polynomials 8.3.17, $g(X_1, X_2, \ldots, X_n)$ is equal to $h(e_1, e_2, \ldots, e_n)$, where h is a polynomial with coefficients in $\mathbb{F}$. It follows that

$$g(\alpha_1, \alpha_2, \ldots, \alpha_n) = h(\beta_1, \beta_2, \ldots, \beta_n)$$

where $\beta_i = e_i(\alpha_1, \alpha_2, \ldots, \alpha_n)$, for $i = 1, 2, \ldots, n$. But, by Remark 8.3.15, each of the β's is in $\mathbb{F}$ (being the quotient of two of the coefficients of f) and thus $g(\alpha_1, \alpha_2, \ldots, \alpha_n)$ is in $\mathbb{F}$. ■

The next proposition will play a vital role in our proof later that π is transcendental.

8.3.24 Proposition. *Let $\mathbb{F}$ be a field and let $t(X)$ be a polynomial of degree n with coefficients in $\mathbb{F}$ and with n zeros $\alpha_1, \alpha_2, \ldots, \alpha_n$ in some extension field $\mathbb{E}$ of $\mathbb{F}$. Assume that k is an integer between 1 and n, and let*

$$\gamma_1, \gamma_2, \ldots, \gamma_m$$

be all the sums of exactly k of the α's. Then there is a monic polynomial $t_k(X)$ of degree m with coefficients in $\mathbb{F}$ which has $\gamma_1, \gamma_2, \ldots, \gamma_m$ as its zeros.

8.3.25 Example. Let $n = 3$ and $k = 2$. Then there are three γ's, namely

$$\gamma_1 = \alpha_1 + \alpha_2, \quad \gamma_2 = \alpha_1 + \alpha_3, \quad \text{and} \quad \gamma_3 = \alpha_2 + \alpha_3.$$

It is clear that $t_2(X)$ must be

$$\begin{aligned} t_2(X) &= (X - \gamma_1)(X - \gamma_2)(X - \gamma_3) \\ &= X^3 - (\gamma_1 + \gamma_2 + \gamma_3)X^2 + (\gamma_1\gamma_2 + \gamma_1\gamma_3 + \gamma_2\gamma_3)X - \gamma_1\gamma_2\gamma_3. \end{aligned}$$

Each coefficient can be expressed in terms of the α's; for example, you can check that

$$\gamma_1\gamma_2 + \gamma_1\gamma_3 + \gamma_2\gamma_3 = \alpha_1^2 + \alpha_2^2 + \alpha_3^2 + 3(\alpha_1\alpha_2 + \alpha_1\alpha_3 + \alpha_2\alpha_3).$$

If you repeat this for other coefficients, you will see that each of the coefficients is a symmetric polynomial (with coefficients which are in $\mathbb{F}$ since they are integers) in the α's, and so is in $\mathbb{F}$ by Corollary 8.3.23. This is the essence of the proof of Proposition 8.3.24. ■

Proof of Proposition 8.3.24. It is clear from the properties required of $t_k(X)$ that we must define it by

$$t_k(X) = (X - \gamma_1)(X - \gamma_2) \ldots (X - \gamma_m).$$

We must prove that each of its coefficients is in $\mathbb{F}$. Now, as in Remark 8.3.15, each of the coefficients of $t_k(X)$ is a symmetric polynomial evaluated at $(\gamma_1, \gamma_2, \ldots, \gamma_m)$. Thus it is sufficient to prove that, if h is any symmetric polynomial in m indeterminates and with coefficients in $\mathbb{F}$ then $h(\gamma_1, \gamma_2, \ldots, \gamma_m) \in \mathbb{F}$.

Accordingly, let h be any symmetric polynomial in m indeterminates and with coefficients in $\mathbb{F}$. We introduce n indeterminates $X_1, X_2, \ldots, X_n$ and let $Y_1, Y_2, \ldots, Y_m$ denote the m distinct sums of the X's taken k at a time. Then $h(Y_1, Y_2, \ldots, Y_m)$ can be expanded to give a polynomial $g(X_1, X_2, \ldots, X_n)$ in the X's

$$h(Y_1, Y_2, \ldots, Y_m) = g(X_1, X_2, \ldots, X_n).$$

It is easy to see that if we permute the X's we also permute the Y's (since a sum of k X's remains such a sum after permutation of the X's, and every such sum comes from another such sum after permutation). Thus $g(X_1, X_2, \ldots, X_n)$ is a symmetric polynomial in $X_1, X_2, \ldots, X_n$ and has its coefficients in $\mathbb{F}$. But, from the connection between the Y's and X's and between the γ's and α's, we have

$$h(\gamma_1, \gamma_2, \ldots, \gamma_m) = g(\alpha_1, \alpha_2, \ldots, \alpha_n)$$

which is in $\mathbb{F}$ by Corollary 8.3.23. ■

Exercises 8.3

1. Write down formulae for the 6 permutations $\rho_1, \rho_2, \ldots, \rho_6$ of $\{X_1, X_2, X_3\}$. If $f(X_1, X_2, X_3) = X_1^2X_2 + X_2^2X_3$, write down $f(\rho_i(X_1, X_2, X_3))$ for $i = 1, 2, \ldots, 6$.
2. Write down two polynomials of degree 4 in X_1, X_2 which are symmetric and two which are not.
3. Assume that $f(X) = 2X^3 - 3X^2 + 4X - 1$ has three zeros $\alpha_1, \alpha_2, \alpha_3$ in $\mathbb{C}$. Calculate
$$\alpha_1 + \alpha_2 + \alpha_3, \ \alpha_1\alpha_2 + \alpha_1\alpha_3 + \alpha_2\alpha_3, \ \alpha_1\alpha_2\alpha_3.$$
4. Apply Algorithm 8.3.19 to express each of the following symmetric polynomials as polynomials in the relevant elementary symmetric polynomials.
   (a) $3X_1^3X_2 + 6X_1^2 - 4X_1^3X_2^3 + 12X_1X_2 + 6X_2^2 + 3X_1X_2^3$.
   (b) $5X_1^2X_2^2X_3 - 2X_1^2X_2X_3 - 2X_1X_2^2X_3 + 5X_1X_2^2X_3^2 + 5X_1^2X_2X_3^2 - 2X_1X_2X_3^2$.
5. Complete Example 8.3.25 by expressing the other coefficients of $t_2(X)$ in terms of the α's and checking that they are symmetric in the α's.
6. Consider the proof of Proposition 8.3.24 in the case $n = 4$, $k = 2$.
   (i) Write down the six γ's.
   (ii) Write down the coefficient of the X^5 term in
   $$t_2(X) = (X - \gamma_1)(X - \gamma_2)\ldots(X - \gamma_6).$$
   Express this in terms of the α's and check it is a symmetric polynomial of the α's, as claimed in the proof of Proposition 8.3.24.
7. (a) Prove (i) of Lemma 8.3.14.

   [Hint. If ρ is any permutation of $\{X_1, \ldots, X_n\}$ then the result of applying ρ to $f(X_1, \ldots, X_n) + g(X_1, \ldots, X_n)$ is $f(\rho(X_1, \ldots, X_n)) + g(\rho(X_1, \ldots, X_n))$.]
   (b) Prove (ii) of Lemma 8.3.14.

   [Hint. If $t(X_1, \ldots, X_n) = h\,(g_1(X_1, \ldots, X_n), \ldots, g_m(X_1, \ldots, X_n))$ then, for any permutation ρ of $\{X_1, \ldots, X_n\}$,

$$t(\rho(X_1, \dots, X_n)) = h(g_1(\rho(X_1, \dots, X_n), \dots, g_m(\rho(X_1, \dots, X_n))).]$$

(c) Give an alternative proof of (ii) of Lemma 8.3.14, this time using (i) of Lemma 8.3.14?

8. A polynomial $f(X_1, X_2, \dots, X_n)$ in $n \in \mathbb{N}$ indeterminates is said to be an *antisymmetric polynomial* if interchanging any two of the indeterminates results in the negative of the polynomial; that is, for $i \neq j$, where $i, j \in \{1, 2, \dots, n\}$,

$$\begin{aligned} &f(X_1, X_2, \dots, X_{i-1}, X_i, \dots, X_{j-1}, X_j, \dots X_n) \\ &= -f(X_1, X_2, \dots, X_{i-1}, X_j, \dots, X_{j-1}, X_i, \dots X_n). \end{aligned}$$

If $f(X_1, X_2)$ is an antisymmetric polynomial, prove that

$$f(X_1, X_2) = (X_1 - X_2)g(X_1, X_2),$$

where $g(X_1, X_2)$ is a symmetric polynomial.

Additional Reading for Chapter 8

An approach to symmetric polynomials which is similar to ours is contained in Ferrar (1958) and Hermite (1873). A proof of the Fundamental Theorem on Symmetric Polynomials which uses a double induction, rather than the concept of order of a monomial, is given in Archbold (1970) and Clark (1971). For further discussion of symmetric polynomials see Lang (2002). The text Macdonald (2015) on Symmetric Polynomials and Hall Polynomials led to what has come to be known as Macdonald polynomials which are central to important developments in the 21st century in mathematical physics, including string theory.

The book Fine and Rosenberger (1997) has a summary of the known proofs (prior to 1997) of The Fundamental Theorem of Algebra 8.2.1. The paper Shipman (2007) has a different proof of an improvement of the Fundamental Theorem of Algebra.

Chapter 9
Quintic Equations II: The Abel–Ruffini Theorem

In this chapter we prove the Abel–Ruffini Theorem which tells us that not all quintic polynomials are algebraically soluble, thereby solving Problem IV in the Introduction. We begin by defining what we mean by algebraically soluble polynomials, proving the Eisenstein Irreducibility Criterion for testing the irreducibility of polynomials and Sturm's Theorem of calculus for determining the number of zeros of a polynomial which are real numbers. This allows us to prove Kronecker's Theorem which provides a sufficient condition for a polynomial being algebraically soluble by counting the number of real number zeros of that polynomial. This leads us immediately to the Abel–Ruffini Theorem.

9.1 Algebraically Soluble Polynomials

9.1.1 Remark. Our primary aim in this chapter is to show that there exists a polynomial $f(X) \in \mathbb{Z}(X)$ of degree 5 which has no zeros which can be obtained from $\mathbb{Q}$ using radicals.
We need to make more explicit what we mean.
As we seek but one example, there is no loss in making restrictions on $f(X)$ as long as we find an example with the required property. Our task will be to find an $f(X)$ which is not algebraically soluble, where this term algebraically soluble is defined next. ■

9.1.2 Definition. Let $f(X) \in \mathbb{Z}(X)$ be a polynomial over $\mathbb{Q}$ of degree n, where n is a positive integer. Then $f(X)$ is said to be *algebraically soluble* if there exist complex numbers $\alpha_1, \alpha_2, \ldots, \alpha_k$ and prime numbers $a_1, a_2, \ldots, a_k$, such that

(i) $\alpha_1^{a_1} \in \mathbb{Q}, \quad \alpha_1 \notin \mathbb{Q}$,
(ii) $\alpha_2^{a_2} \in \mathbb{Q}(\alpha_1), \quad \alpha_2 \notin \mathbb{Q}(\alpha_1)$,

S. A. Morris et al., *Abstract Algebra and Famous Impossibilities*, Undergraduate Texts in Mathematics, https://doi.org/10.1007/978-3-031-05698-7_9

(iii) $\alpha_3^{a_3} \in \mathbb{Q}(\alpha_1, \alpha_2), \quad \alpha_3 \notin \mathbb{Q}(\alpha_1, \alpha_2),$

$\vdots$

(k) $\alpha_k^{a_k} \in \mathbb{Q}(\alpha_1, \alpha_2, \ldots, \alpha_{k-1}), \quad \alpha_k \notin \mathbb{Q}(\alpha_1, \alpha_2, \ldots, \alpha_{k-1})$
such that $f(X)$ is irreducible over $\mathbb{Q}(\alpha_1, \alpha_2, \ldots, \alpha_{k-1})$ but $f(X)$ is reducible over $\mathbb{Q}(\alpha_1, \alpha_2, \ldots, \alpha_k)$. ■

We shall focus on when $f(X)$ is irreducible over $\mathbb{Z}$ and $n = 5$.

9.1.3 Remarks. Note that if $\omega \in \mathbb{Q}(\alpha_1, \alpha_2, \ldots, \alpha_k)$ is a zero of $f(X)$, then $X - \omega$ divides $f(X)$ and so $f(X)$ is indeed reducible over $\mathbb{Q}(\alpha_1, \alpha_2, \ldots, \alpha_k)$. So if $f(X)$ has a zero in radicals, such as $\sqrt[5]{25 + \sqrt[3]{7} + \sqrt{10}}$, then $f(X)$ will be algebraically soluble. Thus if we find a polynomial $f(X)$ which is not algebraically soluble, then we shall know that $f(X)$ has no zeros in radicals.

We should also observe that if instead of prime numbers $a_1, a_2, \ldots, a_k$ in the above definition, we used positive integers greater than 1, the definition of algebraically soluble would be unchanged as it can be reduced easily to the prime number case by factoring each a_i into its prime factors. ■

9.1.4 Remark. Let $z = a + i\,b$ be any complex number and $\overline{z}$ denote its conjugate $a - i\,b$. If z_1 and z_2 are any complex numbers, then it is easily seen that $\overline{z_1 z_2} = (\overline{z_1})(\overline{z_2})$. It follows that if $f(X)$ is any polynomial, then $f(\overline{X}) = \overline{f(X)}$. So if $\alpha \in \mathbb{C}$ is any zero of a polynomial $f(X)$ [that is, $f(\alpha) = 0$], then $f(\overline{\alpha}) = \overline{0} = 0$. Thus if α is a zero of $f(X)$, then so too is its conjugate $\overline{\alpha}$. Thus zeros of $f(X)$ which are not real numbers come in pairs! This is a very useful observation. ■

The Fundamental Theorem of Algebra 8.2.1 tells us that each polynomial of degree n has a zero in $\mathbb{C}$. Indeed, Theorem 8.2.2 says that it has n zeros in $\mathbb{C}$ (some of which may be equal). This observation together with Remark 9.1.4 yields the following two corollaries.

9.1.5 Corollary. *A polynomial $\in \mathbb{Z}[X]$ of degree* **n**, *where* **n** *is an odd integer, must have at least one zero which is a real number.* ■

9.1.6 Corollary. *A polynomial $\in \mathbb{Z}[X]$ of degree 5 must have one, three, or five zeros which are real numbers.* ■

9.1.7 Remark. In due course we shall prove Kronecker's Theorem 9.3.4 which says, in particular, that

if an irreducible polynomial $f(X)$ of degree 5 has three zeros which are real numbers and two zeros which are not real numbers, then $f(X)$ is not algebraically soluble.

Before proving Kronecker's Theorem 9.3.4, we shall see how to produce irreducible polynomials of degree 5 which have precisely three roots which are real numbers. ■

Exercises 9.1

1. Verify that Corollary 9.1.5 would be false if n is not an odd integer by finding a polynomial $f(X) \in \mathbb{Z}[X]$ which has no zeros which are real numbers.
2. Find polynomials of degree 5 which respectively have precisely one, three and five zeros which are real numbers.
3. Extending Exercise 1 above, verify that for every even positive integer n there are an infinite number of polynomials having no real number zeros.

9.2 The Number of Real Number Zeros of an Irreducible Polynomial

In Exercises 6.2 #8 we introduced a simple condition for a polynomial to be irreducible. This condition is sufficient but not necessary. It is generally known today as Eisenstein's Irreducibility Criterion after the German mathematician Ferdinand Gotthold Max Eisenstein (1823–1852) who proved it in Eisenstein (1850). At age 20 he met William Rowan Hamilton who gave him a copy of his book with Niels Henrik Abel's proof of the impossibility of a general solution in radicals of quintic equations and this stimulated his interest in mathematical research. However four years earlier the German mathematician Theodor Schönemann (1812–1868) had already published a stronger result than Eisenstein in Schönemann (1846). For an explanation of why it is called Eisenstein's Irreducibility Criterion (whereas at the beginning of the 20th century it was called Schönemann's Irreducibility Criterion) see Cox (2011). We shall refer to the stronger result as Schönemann's Irreducibility Criterion and the weaker one, which is in many Abstract Algebra textbooks, as the Eisenstein Irreducibility Criterion. There are today many generalizations of these results, including the 1906 work of the Swiss mathematician Gustave Dumas (1872–1955). The Dumas Criterion has the Eisenstein Irreducibility Criterion as a simple corollary. (See Prasolov (2004).)

9.2.1 Theorem. [Schönemann's Irreducibility Criterion] *Let $f(X) \in \mathbb{Z}[X]$ have degree $n > 0$, where $n \in \mathbb{N}$, p is a prime number, and a is an integer such that*

$$f(X) = (X - a)^n + pF(X), \quad F(X) \in \mathbb{Z}[X].$$

If $F(a) \neq 0$, (mod p), then $f(X)$ is irreducible mod p^2.

Proof. Suppose $f(X)$ has a non-trivial factorization mod p^2; that is,

$$f(X) = (X - a)^n + pF(X) = G(X)H(X) \pmod{p^2} \tag{1}$$

where $G(X)$ and $H(X)$ are in $\mathbb{Z}[X]$ and each has degree less than n. So with $i, j > 0$ and $i + j = n$, we have

$$F(X) = f_n X^n + f_{n-1}X^{n-1} + \cdots + f_0$$
$$G(X) = g_i X^i + g_{i-1}X^{i-1} + \cdots + g_0$$
$$H(X) = h_j X^j + h_{j-1}X^{j-1} + \cdots + h_0$$

Balancing coefficients of X^n in (1), we see that $1 + pf_n = g_i h_j \pmod{p^2}$. As $1 \neq 0 \pmod p$, $p \not| g_i$ and $p \not| h_j$.

Now (1) implies that

$$(X - a)^n = G(X)H(X) \pmod p,$$

where the degree of each of the polynomials $G(X)$ and $H(X)$ is greater than or equal to 1 as $p \not| g_i$ and $p \not| h_j$.

If α is a zero of $G(X)$ or $H(X)$, then $G(\alpha) = 0$ or $H(\alpha) = 0$ and so $(\alpha - a)^n = 0 \pmod p$. Thus $p|(\alpha - a)^n$ and since p is prime, $p|(\alpha - a)$; that is $\alpha = a \pmod p$. Hence

$$G(X) = g_i(X - a)^i \pmod p \text{ and } H(X) = h_j(X - a)^j \pmod p.$$

Putting $X = a$ shows that $p|G(a)$ and $p|H(a)$.

Finally, using this and putting $X = a$ in (1), gives $pF(a) = 0 \pmod{p^2}$, which is a contradiction. So our supposition that $f(X)$ has a non-trivial factorization is false and therefore $f(X)$ is irreducible (mod p^2). ■

9.2.2 Corollary. [Eisenstein's Irreducibility Criterion] *Let p be a prime number and let $f(X) = a_n X^n + a_{n-1}X^{n-1} + \cdots + a_1 X + a_0$ be a polynomial in $\mathbb{Z}[X]$ such that*

(i) $a_n \neq 0 \pmod p$,
(ii) $a_{n-1} = a_{n-2} = \cdots = a_0 = 0 \pmod p$, *and*
(iii) $a_0 \neq 0 \pmod{p^2}$.

Then $f(X)$ is irreducible over $\mathbb{Q}$.

Proof. As $a_n \neq 0 \pmod p$ we can multiply by a suitable integer to obtain $f(X) = mg(X), m \in \{1, 2, \ldots, p - 1\}$, so that $g(X)$ is a monic polynomial mod p, and then observe that $f(X)$ is irreducible over $\mathbb{Z}$ if and only if $g(X)$ is. So we can write $g(X) = X^n + pF(X)$, with $F(X) \in \mathbb{Z}[X]$. Note also that $F(0) \neq 0 \pmod p$. So by Schönemann's Irreducibility Criterion 9.2.1, $g(X)$ is irreducible mod p^2. Thus $g(X)$ is irreducible over $\mathbb{Z}$, and so $f(X)$ is irreducible over $\mathbb{Z}$. Hence, by Proposition 8.3.8, Gauss Lemma 2, $f(X)$ is irreducible over $\mathbb{Q}$. ■

9.2.3 Remark. In 1801 Gauss published his major work "Disquisitiones Arithmeticae", Gauss (1966, Latin version appeared in 1801). In §50 of this work was his result that

each cyclotomic polynomial $\frac{X^p-1}{X-1} = X^{p-1} + X^{p-2} + \cdots + 1$, for p a prime number, is irreducible over $\mathbb{Q}$.

To see that this follows easily from Schönemann's Irreducibility Criterion 9.2.1 observe that
$$X^{p-1} + X^{p-2} + \cdots + 1 = (X-1)^{p-1} + pF(X)$$

and that for $X = 1$ we obtain $p = pF(1)$ and so $F(1) = 1 \neq 0 \pmod p$. So by Schönemann's Irreducibility Criterion 9.2.1, $\frac{X^p-1}{X-1}$ is irreducible over p^2, and therefore irreducible over $\mathbb{Q}$. Eisenstein also proved this cyclotomic result but he cannot apply his Criterion immediately, he needs a change of variable $Y = X + 1$. In fact it is the norm that in applying Eisenstein's Irreducibility Criterion 9.2.2 one has to make a change of variable. ■

9.2.4 Proposition. *Let $f(X), g(X) \in \mathbb{Q}[X]$ be irreducible polynomials over $\mathbb{Q}$. Further, let α be a zero of $g(X)$ such that $f(X)$ is a reducible polynomial over the extension field $\mathbb{Q}(\alpha)$. If the degree of $f(X)$ is the prime number p and the degree of $g(X)$ is q, then p is a divisor of q.*

Proof. As $f(X)$ is a reducible polynomial over the extension field $\mathbb{Q}(\alpha)$, we see that $f(X) = \phi(X, \alpha).\psi(X, \alpha)$, where $\phi(X, \alpha)$ and $\psi(X, \alpha)$ are polynomials respectively of degree m and n in $\mathbb{Q}(\alpha)[X]$. Then

$$u(X) = f(r) - \phi(X, r)\psi(X, r), \quad \text{where } r \in \mathbb{Q},$$

vanishes for $X = \alpha$. As $u(X) \in \mathbb{Q}[X]$, by Abel's Irreducibility Theorem 8.3.9, all zeros of $g(X)$ are zeros of $u(X)$, that is, $u(X)$ vanishes at each of these zeros of $g(X)$.

Let α_i be any other zero of $g(X)$. Then

$$f(X) - \phi(X, \alpha_i)\psi(X, \alpha_i) = 0; \text{ that is, } f(X) = \phi(X, \alpha_i)\psi(X, \alpha_i)$$

holds for every rational number X. This says that the polynomial in X on the left hand side equals the polynomial on the right hand side at an infinite number of points, and therefore are equal for every value of $X \in \mathbb{C}$. And this is true for all zeros of $g(X)$.

$$f(X) = \phi(X, \alpha_i)\psi(X, \alpha_i), \text{ for all zeros } \alpha_i \text{ of } g(X).$$

As there are q roots of $f(X)$, multiply all q equations like that above, to obtain
$$f(X)^q = \Phi(X)\Psi(X) \tag{2}$$

where $\Phi(x)$ is the product of all $\phi(X, \alpha_i)$ and $\Psi(X)$ is the product of all $\psi(X, \alpha_i)$ as α_i runs over the q roots of $g(X)$. Since $\Psi(X)$ and $\Phi(X)$ are symmetric polynomials of the roots of $g(X)$, by The Fundamental Theorem on Symmetric Polynomials 8.3.17, $\Phi(X), \Psi(X) \in \mathbb{Q}[X]$.

Noting that the degree of the polynomial $f(X)$ is greater than that of $\phi(x, \alpha)$ and that of $\psi(x, \alpha)$, Theorem 8.2.2 implies that at least one zero of $f(X)$ must be a zero of $\phi(x, \alpha)$ and at least one zero of $f(x)$ must be a zero of $\psi(x, \alpha)$. Consequently, at least one zero of $f(x)$ is a zero of $\Psi(x)$. The same is true for $\Phi(X)$. By Abel's

Irreducibility Theorem 8.3.9 every zero of $f(X)$ is a zero of $\Phi(X)$ and of $\Psi(X)$. Indeed $\Phi(X) = f(X)G_1(X)$, $G_1(X) \in \mathbb{Q}[x]$.

By (2), every zero of $G_1(X)$ is then a zero of $f(X)^q$, and hence of $f(X)$. So $G_1(X) = f(X)G_2(X)$, where the degree of the polynomial $G_2(X) \in \mathbb{Q}(X)$ is strictly less than that of $G_1(X)$. Continuing in this way we obtain $\Phi(X) = f(X)^\mu$, for some positive integer μ. Similarly $\Psi(X) = f(X)^\nu$, for some positive integer ν. So $q = \mu + \nu$.

Noting that by the assumption that the degree of the polynomial $f(X)$ is p and the degree of each $\phi(X, \alpha)$ is m, we have that $mq = \mu p$. Similarly $nq = \nu p$. As m and n are smaller than p, it follows that p is a divisor of q, as required. ■

9.2.5 Corollary. *Let $f(X) \in \mathbb{Q}[X]$ be an irreducible polynomial over $\mathbb{Q}$ of degree p, for p a prime number. Let $\alpha \neq 1$ be a qth complex number root of unity (i.e. $\alpha^q = 1$), where q is a prime number and $q \leq p$. Then $f(X)$ is also an irreducible polynomial over $\mathbb{Q}(\alpha)$.*

Proof. By Remark 9.2.3 α is a zero of the cyclotomic polynomial

$$X^{q-1} + X^{q-2} + \cdots + 1$$

which is irreducible. Suppose $f(X)$ is reducible over $\mathbb{Q}(\alpha)$. Then by Proposition 9.2.4, p would divide $q - 1$, which is false. Therefore, our supposition is false and $f(X)$ is also irreducible over $\mathbb{Q}(\alpha)$. ■

9.2.6 Example. Let $f(X) = X^5 - 10X - 5$.

By Eisenstein's Irreducibility Criterion 9.2.2, $f(X)$ is irreducible over $\mathbb{Q}$.

Firstly, let us find the stationary points of f, that is where $\frac{df(X)}{dX} = f'(X) = 0$. As $f'(X) = 5X^4 - 10$, the stationary points are $X = -\sqrt[4]{2}$ and $X = \sqrt[4]{2}$ (Fig. 9.1).

In fact, we see that $f'(X)$ is positive for $X < -\sqrt[4]{2}$, negative for $-\sqrt[4]{2} < X < \sqrt[4]{2}$, and positive for $X > \sqrt[4]{2}$; thus f is an increasing function for $x < -\sqrt[4]{2}$, a decreasing function for $-\sqrt[4]{2} < X < \sqrt[4]{2}$, and an increasing function for $X > \sqrt[4]{2}$.

Clearly $f(X) \to \infty$ as $X \to \infty$ and $f(X) \to -\infty$ as $X \to -\infty$. [Indeed $f(-2) < 0$, $f(-1) > 0$, $f(0) = -5$, $f(1) < 0$, and $f(2) > 0$.]

Thus the graph of $f(X)$ must cross the X-axis precisely 3 times and look like Fig. 9.1.

This suggests that $f(X)$ has precisely 3 zeros which are real numbers. But we must be careful to check that none of these are repeated zeros, as we need to know that there are zeros which are not real numbers.

If α is a repeated zero of $f(X)$, then

$$f(X) = (X - \alpha)(X - \alpha)(X - \beta)(X - \gamma)(X - \delta), \text{ for } \alpha, \beta, \gamma, \delta \in \mathbb{C};$$

that is, $f(X) = (X - \alpha)[(X - \alpha)g(X)]$, where $g(X) = (X - \beta)(X - \gamma)(X - \delta)$. Then $f'(X) = 1.[(X - \alpha)g(X)] + (X - \alpha)[(X - \alpha)g(X)]'$. Hence $f'(\alpha) = 0$. So if α is a repeated zero of $f(X)$, then α is also a stationary point of f.

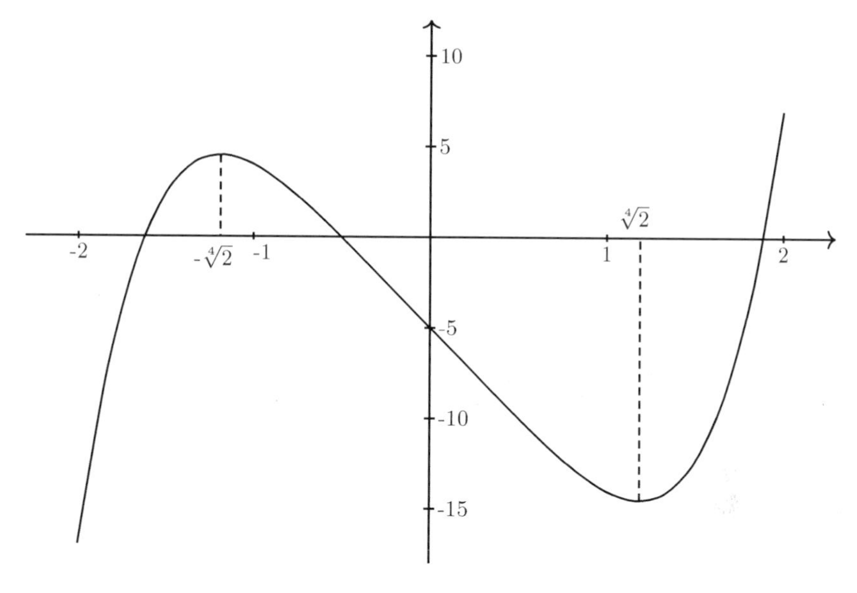

Fig. 9.1 $f(X) = X^5 - 10X - 5$

As neither of the stationary points of f is a zero of $f(X)$, we have proved that $f(X) = X^5 - 10X - 5$ has 3 zeros which are real numbers and 2 zeros which are not real numbers. ■

9.2.7 Remark. The problem of finding the number of real number zeros of a polynomial was solved in Sturm (1829) by the French-Swiss mathematician Jacques Charles François Sturm (1803–1855). The idea is a generalization of the technique used above. Without going into detail, one associates with each polynomial a sequence of polynomials, called the *Sturm sequence* and the number of real number roots is determined by an examination of the sign changes of the polynomials in the Sturm sequence. While the Fundamental Theorem of Algebra 8.2.1 tells us the number of complex number zeros, Sturm's sequence determines the number of distinct real number zeros and locates an interval containing each one. Indeed by subdividing these intervals, one can approximate these roots as closely as one wishes. ■

9.2.8 Remark. Using the same technique as in Example 9.2.6 we can readily obtain the following proposition which generalizes that example and yields an infinite number of irreducible quintic polynomials with three real number zeros and two zeros which are not real numbers. Such polynomials by Kronecker's Theorem 9.3.4 will therefore be not algebraically soluble. ■

The details of the proof of the next proposition are left as an exercise.

9.2.9 Proposition. *Let $f(X) \in \mathbb{Z}[X]$ be given by $f(X) = X^5 - aX - b$ where*
(i) p is a prime number;

(ii) a and b are positive integers divisible by p;
(iii) b is not divisible by p^2; and
(iv) $a^5 > \frac{5^5}{4^4} b^4$.
Then $f(X)$ is an irreducible polynomial which has precisely 3 zeros which are real numbers and 2 zeros which are not real numbers.

There is an infinite number of such $f X$).

9.2.10 Remark. Note that condition (iv) in Proposition 9.2.9 is satisfied if $a \geq 2b$. So the irreducible polynomial $f(X) = X^5 - 10X - 5$ has precisely 3 zeros which are real numbers and 2 zeros which are not real numbers. ■

As a generalization of Proposition 9.2.9, we note the following Proposition. For a proof similar to that of Proposition 9.2.9 but using Sturm's Theorem, see Dörrie (1965).

9.2.11 Proposition. *If $g(X)$ is the polynomial given by $g(X) = X^n - aX - b$ where*
(1) n is an odd positive integer ≥ 5;
(ii) p is a prime number;
(iii) a and b are positive integers divisible by p;
(iv) b is not divisible by p^2; and
(v) $a^n > \frac{n^n}{(n-1)^{(n-1)}} b^{n-1}$.
Then $g(X)$ is an irreducible polynomial which has precisely 3 zeros which are real numbers and $n - 3$ zeros which are not real numbers. Indeed, for each n, there is an infinite number of such $g(X)$.

We conclude this section with Abel's Lemma.

9.2.12 Proposition. [Abel's Lemma] *Let p be a prime number, $\mathbb{F}$ a subfield of $\mathbb{C}$, $c \in \mathbb{F}$ but $c \neq d^p$, for any $d \in \mathbb{F}$. Then the polynomial $X^p - c$ is irreducible over $\mathbb{F}$.*

Proof. Suppose $X^p - c$ is reducible over $\mathbb{F}$; that is

$$X^p - c = \psi(X)\phi(X)$$

where $\psi(X), \phi(X) \in \mathbb{F}[X]$. Let the constant terms of $\psi(X)$ and $\phi(X)$ be a and b, respectively, where $a, b \in \mathbb{F}$. Since the zeros of $X^p - c$ are $r, r\varepsilon, r\varepsilon^2, \ldots, r\varepsilon^{p-1}$, where r is one of the zeros and ε is a complex p^{th} root of unity. Then $a = r^\mu \varepsilon^M$ and $b = r^\nu \varepsilon^N$, for some $M, N \in \mathbb{N}$. Since $\nu + \mu = p$, where p is a prime number, ν and μ have no common factor. Thus there exist integers h and k such that $\mu h + \nu k = 1$. Putting $K = a^h b^k$, we see that $K = r\varepsilon^{hM+kN}$. Thus

$$K^p = r^p = c.$$

This says that the $K \in \mathbb{F}$ is a pth root of c, which is a contradiction. So our supposition is false, and $X^p - c$ is irreducible over $\mathbb{F}$. ■

Exercises 9.2

1. Using Eisenstein's Irreducibility Criterion 9.2.2 prove that the polynomial $f(X) = 3X^4 + 15X^2 + 10$ is irreducible over $\mathbb{Q}$.
2. Using Remark 9.2.3 and Eisenstein's Irreducibility Criterion 9.2.2 prove that the cyclotomic polynomial is irreducible over $\mathbb{Q}$.
3. Let n be any positive integer greater than 3. Use Eisenstein's Irreducibility Criterion to prove that $f(X) = X^n + 10X^2 + 15X + 5$ is irreducible over $\mathbb{Q}$.
4. Let $f(X) = X^{2107} - 30X - 20 \in \mathbb{Z}[X]$. Verify that $f(X)$ is irreducible over $\mathbb{Q}$ and determine how many zeros of $f(X)$ are real numbers and how many zeros are not real numbers.

9.3 Kronecker's Theorem and the Abel–Ruffini Theorem

The following beautiful theorem, which appears not to be well-known, was proved by the German mathematician Leopold Kronecker (1823–1891) in 1858. (He is credited with the remark: "Die ganzen Zahlen hat der liebe Gott gemacht, alles andere ist Menschenwerk" ("God made the integers, all else is the work of man.") Our proof of Kronecker's Theorem follows that by the German mathematician Heinrich Dörrie (1873–1955) in (Dörrie, 1965, p. 127, §25) except for a small modification to avoid a gap in the proof in Dörrie (1965) identified in Pan and Chen (2020). Kronecker's Theorem 9.3.4 is also an immediate consequence of a result of Galois.

9.3.1 Definition. Let $\mathbb{F}$ be a subfield of $\mathbb{C}$. If for each $a \in \mathbb{F}$, the complex conjugate $\overline{a}$ is in $\mathbb{F}$, then $\mathbb{F}$ is said to be a *complex conjugate closed field.* ■

9.3.2 Examples. The following are examples of complex conjugate closed fields:

(i) $\mathbb{Q}$;
(ii) $\mathbb{R}$;
(iii) all subfields of $\mathbb{R}$;
(iv) $\mathbb{C}$;
(iv) $\mathbb{Q}(a, \overline{a})$ for any $a \in \mathbb{C}$;
(v) $\mathbb{Q}(a_1, a_2, \ldots, a_n, \overline{a_1}, \overline{a_2}, \ldots, \overline{a_n})$, for any $a_1, a_2, \ldots, a_n \in \mathbb{C}$, and $n \in \mathbb{N}$. ■

9.3.3 Proposition. *Let $f(X) \in \mathbb{Z}[X]$ be a monic polynomial of degree p, where p is an odd prime number such that $f(X)$ is algebraically soluble. Let $b_1 = p$ and β_1 a pth root of unity, where $\beta_1 \notin \mathbb{Q}$. Then there exists complex numbers $\beta_2, \ldots, \beta_k$ and prime numbers $b_2, \ldots, b_k$ such that*
(i) $\beta_1^{b_1} = 1 \in \mathbb{Q}$, *where* $\beta_1 \notin \mathbb{Q}$;
(ii) $\beta_2^{b_2} \in \mathbb{Q}(\beta_1, \overline{\beta_1})$;
(iii) $\beta_3^{b_3} \in \mathbb{Q}(\beta_1, \overline{\beta_1}, \beta_2, \overline{\beta_2})$;

$\vdots$

(k) $\beta_k^{b_k} \in \mathbb{Q}(\beta_1, \overline{\beta_1}, \beta_2, \overline{\beta_2}, \ldots, \beta_{k-1}, \overline{\beta_{k-1}})$,

and for $k > 1$, the polynomial $f(X)$ is irreducible over the fields $\mathbb{Q}(\beta_1, \overline{\beta_2})$, $\mathbb{Q}(\beta_1, \overline{\beta_1}, \beta_2, \overline{\beta_2}), \ldots, \mathbb{Q}(\beta_1, \overline{\beta_1}, \beta_2, \overline{\beta_2}, \ldots, \beta_{k-1}, \overline{\beta_{k-1}})$ and $f(X)$ is reducible over the field $\mathbb{Q}(\beta_1, \overline{\beta_1}, \beta_2, \overline{\beta_2}, \ldots, \beta_k, \overline{\beta_k})$.
Further, each of the fields $\mathbb{Q}(\beta_1, \overline{\beta_1}, \beta_2, \overline{\beta_2}, \ldots, \beta_n, \overline{\beta_n})$, $n = 1, 2, \ldots, k$, is a complex conjugate closed field.

Proof. As $f(X)$ is algebraically soluble, there exist complex numbers $\alpha_1, \alpha_2, \ldots, \alpha_{k^*}$ and prime numbers $a_1, a_2, \ldots, a_{k^*}$, such that
(i*) $\alpha_1^{a_1} \in \mathbb{Q}, \quad \alpha_1 \notin \mathbb{Q}$,
(ii*) $\alpha_2^{a_2} \in \mathbb{Q}(\alpha_1), \quad \alpha_2 \notin \mathbb{Q}(\alpha_1)$,
(iii*) $\alpha_3^{a_3} \in \mathbb{Q}(\alpha_1, \alpha_2), \quad \alpha_3 \notin \mathbb{Q}(\alpha_1, \alpha_2)$,

$\vdots$

(k*) $\alpha_k{*}^{a_{k*}} \in \mathbb{Q}(\alpha_1, \alpha_2, \ldots, \alpha_{k^*-1}), \quad \alpha_k \notin \mathbb{Q}(\alpha_1, \alpha_2, \ldots, \alpha_{k^*-1})$

and $f(X)$ is irreducible over $\mathbb{Q}(\alpha_1, \alpha_2, \ldots, \alpha_{k^*-1})$ and $f(X)$ is reducible over $\mathbb{Q}(\alpha_1, \alpha_2, \ldots, \alpha_{k^*})$.

By Corollary 9.2.5, without loss of generality we can let $\alpha_1 \neq 1$ be a complex number p^{th} root of unity (so $a_1 = p$), and we have that $f(X)$ is irreducible over $\mathbb{Q}(\alpha_1)$ and so $k^* > 1$. As the complex conjugate of α_1 is a pth root of unity and all p^{th} roots of unity are powers of α_1, it follows that each is in $\mathbb{Q}(\alpha_1)$. So $\mathbb{Q}(\alpha_1) = \mathbb{Q}(\alpha_1, \overline{\alpha_1})$. Thus $f(X)$ is irreducible over the complex conjugate closed field $\mathbb{Q}(\alpha_1, \overline{\alpha_1})$.

Using mathematical induction, as pointed out in Examples 9.3.2(v), it is clear that each of the fields $Q(\beta_1, \overline{\beta_1}, \beta_2, \overline{\beta_2}, \ldots, \beta_n, \overline{\beta_n})$, $n = 1, 2, \ldots, k$, is a complex conjugate closed field.

Let us consider the following:

(a) $\mathbb{Q}$ is a complex conjugate closed field. Put $\beta_1 = \alpha_1$ and $b_1 = a_1$. Then $\beta_1^{b_1} = \alpha_1^{a_1} \in \mathbb{Q}$ and $\beta_1 \neq 1$, is a complex number p^{th} root of unity. So (i) is true.
As noted above, $f(X)$ is irreducible over $\mathbb{Q}(\beta_1, \overline{\beta_1})$.
(b) Put $\beta_2 = \alpha_2$ and $b_2 = a_2$. Then we obtain $\beta_2^{b_2} = \alpha_2^{a_2} \in \mathbb{Q}(\alpha_1) = \mathbb{Q}(\beta_1) \subseteq \mathbb{Q}(\beta_1, \overline{\beta_1})$. So (ii) is true.
Either $f(X)$ is reducible or irreducible over $\mathbb{Q}(\beta_1, \overline{\beta_1}, \beta_2, \overline{\beta_2})$. If it is reducible, then k=2. If it is irreducible, we continue with the process.
(c) We continue the process in the above manner until either the process has stopped or we reach $\beta_{k^*} = \alpha_{k^*}$. In the latter case, as $f(X)$ is algebraically soluble, it is reducible over $\mathbb{Q}(\alpha_1, \alpha_2, \ldots \alpha_{k^*})$ and thus it is reducible over $\mathbb{Q}(\beta_1, \overline{\beta_1}, \beta_2, \overline{\beta_2}, \ldots \beta_{k^*}, \overline{\beta_{k^*}})$ since

$$\mathbb{Q}(\beta_1, \overline{\beta_1}, \beta_2, \overline{\beta_2}, \ldots \beta_{k^*}, \overline{\beta_{k^*}}) \supseteq \mathbb{Q}(\beta_1, \beta_2, \ldots \beta_{k^*}) = \mathbb{Q}(\alpha_1, \alpha_2, \ldots \alpha_{k^*}),$$

which completes the proof of the proposition. ∎

9.3.4 Theorem. [Kronecker's Theorem] *Let $f(X) \in \mathbb{Z}[X]$ be irreducible over $\mathbb{Z}$ (or equivalently over $\mathbb{Q}$) and have degree p, which is an odd prime number. If $f(X)$ is algebraically soluble, then either it has only one zero which is a real number or all of its zeros are real numbers.*

Proof. Let $f(X)$ be algebraically soluble. Then $f(X)$ satisfies the conditions in Proposition 9.3.3. For consistency we now replace β by α and b by a in the statement of Proposition 9.3.3. For convenience we introduce some notation:

(a) Define the fields $\mathbb{K} = \mathbb{Q}(\alpha_1, \alpha_2, \dots, \alpha_{k-1})$ and $\mathbb{L} = \mathbb{Q}(\alpha_1, \alpha_2, \dots, \alpha_{k-1}, \alpha_k)$,
(b) $l = a_k$, $\lambda = \alpha_k$, and $K \in \mathbb{K}$ such that $\lambda = \sqrt[l]{K}$. So

$$f(X) = \phi_1(X, \lambda)\phi_2(X, \lambda) \dots \phi_m(X, \lambda)$$

where $\phi_1, \phi_2, \dots, \phi_m$ are not in $\mathbb{K}[X]$, but are irreducible polynomials in $\mathbb{L}[X]$ with coefficients which are polynomials in λ. So by Proposition 9.2.4 the prime number p must divide the prime number l. So $l = p$.

Put $\eta = \alpha_1$. Then the p roots of $\lambda^p = K$ are

$$\lambda_0 = \lambda,\ \lambda_1 = \lambda\eta,\ \lambda_2 = \lambda\eta^2,\ \dots,\ \lambda_\nu = \lambda\eta^\nu,\ \dots,\ \lambda_{p-1} = \lambda\eta^{p-1}.$$

Since each $\phi_i(X, \lambda)$ divides $f(X)$, it follows from the proof of Proposition 9.2.4 that each $\phi_i(X, \lambda_\nu)$ also divides $f(X)$.

Suppose $\phi_i(X, \lambda_\nu)$ is reducible over $\mathbb{L}$. Then

$$\phi_i(X, \lambda_\nu) = u(X, \lambda_\nu)v(X, \lambda_\nu).$$

Thus $\phi_i(X, \lambda) = u(X, \lambda)v(X, \lambda)$, and using the same method as in Proposition 9.2.4 this gives a contradiction since $\phi_i(X, \lambda)$ is irreducible over $\mathbb{L}$. So each $\phi_i(X, \lambda_\nu)$ is irreducible over $\mathbb{L}$.

Now suppose $\phi_i(X, \lambda_\nu) = \phi_i(X, \lambda_\mu)$, for some $1 \le \mu < \nu \le p - 1$. Putting $\delta = \eta^{\nu-\mu}$, we can replace λ by the zero $\lambda\eta^{\nu-\mu} = \lambda\delta$, from which it follows that

$$\phi_i(X, \lambda) = \phi_i(X, \lambda\delta).$$

Similarly by replacing successively δ by $\lambda\delta, \lambda\delta^2, \dots$ it follows that

$$\phi_i(X, \lambda) = \phi_i(X, \lambda\delta) = \phi_i(X, \lambda\delta^2) = \phi_i(X, \lambda\delta^3) \dots.$$

This yields

$$\phi_i(X, \lambda) = \frac{\phi_i(X, \lambda) + \phi_i(X, \lambda\delta) + \dots + \phi_i(X, \lambda\delta^{p-1})}{n}.$$

The right hand side of this equation is a symmetric polynomial in the zeros λ, $\lambda\delta$, $\lambda\delta^2, \dots$ of the polynomial $X^n - K \in \mathbb{K}[X]$ and so is a polynomial in $\mathbb{K}[X]$. Thus

$\phi_i(X, \lambda)$ is a polynomial in $\mathbb{K}[X]$, which is a contradiction to $f(X)$ being irreducible in $\mathbb{K}[X]$. Therefore our supposition is false, and hence all $\phi_i(X, \lambda_\nu)$ are distinct.

So we see that $f(X)$ is divisible by the product $\Phi(X)$ of p distinct factors $\phi_i(X, \lambda), \phi_i(X, \lambda\eta), \dots, \phi_i(X, \lambda\eta^{p-1})$ that are irreducible in $\mathbb{L}[X]$:

$$f(X) = \Phi(X)U(X),$$

where $\Phi(X)$ is a symmetric polynomial function of the zeros of $X^n - K \in \mathbb{K}[X]$. Consequently Φ and U are polynomials in $\mathbb{K}[X]$. As $f(X)$ is irreducible in $\mathbb{K}[X]$, $U(X)$ must be 1, and so

$$f(X) = \Phi(X) = \phi_i(X, \lambda)\phi_i(X, \lambda\eta) \dots \phi_i(X, \lambda\eta^{p-1}).$$

So we see that the divisibility of $f(X)$ over $\mathbb{L}$ is thus a divisibility into linear factors. If $\omega, \omega_1, \dots, \omega_{p-1}$ are the zeros and $X - \omega, X - \omega_1, \dots, x - \omega_{p-1}$ are the linear factors of $f(X)$, then

$$X - \omega = \phi_i(X, \lambda),\ X - \omega_1 = \phi_i(X, \lambda\eta),\ \dots,\ X - \omega_{p-1} = \phi_i(X, \lambda\eta^{p-1}).$$

Thus

$$\begin{aligned}
\omega &= K_0 + K_1\lambda + K_2\lambda^2 + \dots + K_{p-1}\lambda^{p-1} \\
\omega_1 &= K_0 + K_1\lambda_1 + K_2\lambda_1^2 + \dots + K_{p-1}\lambda_1^{p-1} \\
&\vdots \\
\omega_{p-1} &= K_0 + K_1\lambda_{p-1} + K_2\lambda_{p-1}^2 + \dots + K_{p-1}\lambda_{p-1}^{p-1}
\end{aligned}$$

where all the $K_\nu \in \mathbb{K}$, since we saw earlier that $\phi(X, \lambda)$ has coefficients which are polynomials of λ and $\lambda^p = K \in \mathbb{K}$.

By Corollary 9.1.5, the polynomial $f(X)$ of odd degree must have at least one real zero. Let this zero which is a real number be

$$\omega = K_0 + K_1\lambda + K_2\lambda^2 + \dots + K_{p-1}\lambda^{p-1}.$$

We distinguish two cases:

(1) K is a real number;
(2) K is a complex number which is not a real number.

Firstly we consider Case (1) where K is real. Noting that $\mathbb{K}$ contains the pth roots of unity, we can assume λ is a real number. So the complex conjugate of ω is given by

$$\overline{\omega} = \overline{K_0} + \overline{K_1}\lambda + \dots + \overline{K_{p-1}}\lambda^{p-1},$$

where each complex conjugate $\overline{K_\nu} \in \mathbb{K}$, as $\mathbb{K}$ is a complex conjugate closed field.

As we said that ω is a real number it follows that $\omega - \overline{\omega} = 0$; that is,

$$(\overline{K_0} - K_0) + (\overline{K_1} - K_1)\lambda + \cdots + (\overline{K_{p-1}} - K_{p-1})\lambda^{p-1} = 0$$

which by Proposition 8.3.9 implies that each $\overline{K_\nu} - K_\nu = 0$. Thus $K_0, K_1, \ldots, K_{p-1}$ are all real numbers. Further,

$$\omega_\nu = K_0 + K_1\lambda_\nu + \cdots + K_{p-1}\lambda_\nu^{p-1}$$

and

$$\omega_{p-\nu} = K_0 + K_1\lambda_{p-\nu} + \cdots + K_{p-1}\lambda_{p-\nu}^{p-1}.$$

However, since $\lambda_\nu = \lambda\eta^\nu$ and $\lambda_{p-\nu} = \lambda\eta^{p-\nu} = \lambda\eta^{-\nu}$ are complex conjugates, it follows that ω_ν and $\omega_{p-\nu}$ are also complex conjugates.

So we have shown that in Case (1)

$f(X)$ has precisely one zero which is a real number and $p - 1$ paired conjugate complex number zeros which are not real numbers.

Now we consider Case (2), where $K \in \mathbb{C} \setminus \mathbb{R}$. We know that $f(X)$ is irreducible over $\mathbb{K}$ but is reducible over $\mathbb{K}(\lambda)$, where $\lambda = \sqrt[p]{K}$. Put $\Lambda = \lambda\overline{\lambda} = \sqrt[p]{K\overline{K}}$. If $f(X)$ is reducible over $\mathbb{K}(\Lambda)$, then we are in Case 1 as Λ is a real number. So we may assume that $f(X)$ is irreducible over $\mathbb{K}(\Lambda)$. From

$$\omega = K_0 + K_1\lambda + K_2\lambda^2 + \cdots + K_{p-1}\lambda^{p-1}$$

it follows that

$$\begin{aligned}\overline{\omega} &= \overline{K_0} + \overline{K_1}\overline{\lambda} + \overline{K_2}\,\overline{\lambda}^2 + \cdots + \overline{K_{p-1}}\,\overline{\lambda}^{p-1} \\ &= \overline{K_0} + \overline{K_1}\left(\frac{\Lambda}{\lambda}\right) + \overline{K_2}\left(\frac{\Lambda}{\lambda}\right)^2 + \cdots + \overline{K_{p-1}}\left(\frac{\Lambda}{\lambda}\right)^{p-1}.\end{aligned}$$

Since $\omega = \overline{\omega}$, this implies that

$$\begin{aligned}&K_0 + K_1\lambda + K_2\lambda^2 + \cdots + K_{p-1}\lambda^{p-1} \\ =&\overline{K_0} + \overline{K_1}\left(\frac{\Lambda}{\lambda}\right) + \overline{K_2}\left(\frac{\Lambda}{\lambda}\right)^2 + \cdots + \overline{K_{p-1}}\left(\frac{\Lambda}{\lambda}\right)^{p-1}.\end{aligned}$$

In this equation all quantities with the exception of λ are in $\mathbb{K}(\Lambda)$. By Proposition 9.2.12 (Abel's Lemma), the polynomial $X^p - K$ is irreducible over $\mathbb{K}(\Lambda)$. Therefore we can replace λ in the above equation by any root λ_ν of $X^p - K$. Recalling that

$$\frac{\Lambda}{\lambda_\nu} = \frac{\Lambda}{\lambda\eta^\nu} = \frac{\overline{\lambda}}{\eta^\nu} = \overline{\lambda}\overline{\eta}^\nu = \overline{\lambda\eta^\nu} = \overline{\lambda_\nu}$$

we see that

$$K_0 + K_1\lambda_\nu + \cdots + K_{p-1}\lambda_\nu^{p-1} = \overline{K_0} + \overline{K_1}\,\overline{\lambda_\nu} + \cdots + \overline{K_{p-1}}\,\overline{\lambda_\nu}^{p-1};$$

that is, $\omega_\nu = \overline{\omega_\nu}$. Thus in Case (2)
all of the zeros of $f(X)$ are real numbers. ∎

As an immediate corollary of Kronecker's Theorem 9.3.4 and Proposition 9.2.9 we have the following powerful Theorem. It answered the question of the solvability of quintic equations in radicals which had been investigated for over 250 years by many of the best mathematicians of the time.

9.3.5 Theorem. *There exist an infinite number of irreducible polynomials $f(X) \in \mathbb{Z}[X]$ of degree 5 which are not algebraically soluble.* ∎

If one uses Proposition 9.2.11, which is a generalization of Proposition 9.2.9, we obtain

9.3.6 Theorem. *If $p \geq 5$ is a prime number, then there exist an infinite number of irreducible polynomials $f(X) \in \mathbb{Z}[X]$ of degree p which are not algebraically soluble.*

As mentioned earlier, Kronecker's Theorem 9.3.4 is obviously an immediate consequence of Theorem 9.3.7 of the French mathematician Évariste Galois (1811–1832). Unfortunately the original statement of the theorem of Galois at first glance did not appear to be a breakthrough. But it implies Kronecker's Theorem 9.3.4 and so is indeed a breakthrough. For further discussion, see Edwards (1984); Rosen (1995).

9.3.7 Theorem. [Galois] *Let $f(X)$ be an irreducible polynomial of degree p over $\mathbb{Q}$. If $f(X)$ is algebraically soluble, p is an odd prime number, and $\theta_1, \theta_2, \ldots, \theta_p$ are the zeros of $f(X)$, then $\theta_n \in \mathbb{Q}(\theta_1, \theta_2)$, for $n = 1, 2, \ldots, p$.*

9.3.8 Remark. Recall that the term algebraically soluble was defined in Definition 9.1.2. Theorem 9.3.6 then tells us that for each prime number p greater than or equal to 5, there exist an infinite number of irreducible polynomials $f(X)$ which do not have zeros in terms of rational numbers and radicals. We have seen, in fact, how to find quite easily an infinite number of such irreducible polynomials. (Some textbooks show only that there exists one such polynomial for $p = 5$.)

The "modern" approach to the Abel–Ruffini Theorem 9.3.9, stated below, uses Galois Theory, named after Évariste Galois (For an analysis of the work of Galois, see Neumann (2010) (1811–1832).) This approach is to be found in textbooks on Abstract Algebra such as MacLane and Birkhoff (1988) and Fraleigh (1982). The idea is that a polynomial $f \in \mathbb{Z}[X]$ is algebraically soluble if and only of its Galois group is soluble, (Stewart, 2004, Theorem 18.19). The Galois group of any polynomial $f(X) \in \mathbb{Z}[X]$ is shown to be isomorphic to a group of permutations of the distinct roots of the polynomial. So the Galois group of a so-called "general" polynomial of degree n (which is, in fact, not general in a real sense, see Stewart (2004); Edwards (1984); Lorenz (2006)) is isomorphic to a subgroup of the Symmetric Group $\mathcal{S}_n$ of all permutations of a set with n elements. If p is a prime number, then the Galois group of an irreducible polynomial $f(X)$ of degree p is $\mathcal{S}_p$ itself. (See (Stewart, 2004, Lemma 15.10).) As every subgroup of a soluble group is soluble and $\mathcal{S}_n$ has the Alternating Group $\mathbb{A}_n$ of all even permutations as a subgroup, and $\mathbb{A}_n$ is known to be a simple group for all $n \geq 5$, and simple implies not soluble, the group $\mathcal{S}_n$ is not soluble for any $n \geq 5$. For a thorough presentation of Galois Theory, see Stewart (2004); Edwards (1984); Lorenz (2006).

9.3.9 Theorem. [Abel–Ruffini Theorem] *If the prime number p satisfies $p \geq 5$, then there exist an irreducible polynomial $f(X) \in \mathbb{Z}[X]$ of degree p which is not algebraically soluble.*

9.3.10 Remark. You should note that we have not proved that all irreducible quintic polynomials are not solvable by radicals. Of course some quintic polynomials are solvable by radicals, for example $X^5 - 1$. So it is reasonable to ask which ones are and what are their solutions. There is a characterization of the quintic equations which are solvable by radicals in Spearman and Williams (1994). Further, Dummit (1991) shows how to solve those quintic equations which are solvable.

Exercises 9.3

1. Verify that Kronecker's Theorem 9.3.4 is an immediate consequence of the Galois Theorem 9.3.7.
2. Verify that the following polynomials are not algebraically soluble:

   (i) $f_1(X) = X^5 - 15X - 5$
   (ii) $f_2(X) = X^7 - 14X - 7$.

3. Verify that each of the numbers 1, 2, 3, 4, and 5 are zeros of the polynomial $f(X) = X^5 - 15X^4 + 85X^3 - 225X^2 + 274X - 120$. Observe that this says that $f(X)$ is not irreducible over $\mathbb{Q}$. Verify, however, that $g(X) = f(X) + X$ is an irreducible polynomial over $\mathbb{Q}$.
   Note the following definition.

9.3.2 Definition. A polynomial $f(X) \in \mathbb{F}[X]$ is said to be *solvable by radicals over* $\mathbb{F}$ if there are complex numbers $c_1, c_2, c_3, \ldots, c_n$ and positive integers $i_1, i_2, i_3, \ldots, i_n$ such that

$$\begin{aligned}
{c_1}^{i_1} &\in \mathbb{F} \\
{c_2}^{i_2} &\in \mathbb{F}(c_1) \\
{c_3}^{i_3} &\in \mathbb{F}(c_1)(c_2) \\
&\vdots \\
{c_n}^{i_n} &\in \mathbb{F}(c_1)(c_2)\ldots(c_{n-1})
\end{aligned}$$

and the zeros in $\mathbb{C}$ of $f(X)$ are all contained in the field

$$\mathbb{F}(c_1)(c_2)\ldots(c_{n-1})(c_n)\,. \qquad \blacksquare$$

4. Using Exercise 2 above, prove that *there exist polynomials of every degree* $n \geq 5$ *which are not solvable by radicals over* $\mathbb{Q}$.
   [Hint. Multiply by X^{n-5}, for $n \in \mathbb{N}$.]

Additional Reading for Chapter 9

There are many textbooks on Abstract Algebra such as Fraleigh (1982) and MacLane and Birkhoff (1988) which cover in an elementary way topics in this chapter. There are many books on Galois Theory, and we mention specifically Bewersdorff (2006); Stewart (2004); Cox (2012); Edwards (1984); Rotman (2001); Artin and Milgram (1971); Lorenz (2006) which are suitable introductions. Peter Michael Neumann (1940–2020) has a study in Neumann (2010) of the contributions of Galois. (Clark, 1971, Sections 139–149) gives an extensive account of the application of Galois theory to the solution by radicals of polynomial equations of low degree. A re-examination of the short, but action-packed, life of Galois can be found in Rothman (1982). Dörrie (1965) has a presentation of the Abel–Ruffini Theorem avoiding Galois Theory. The history of the Eisenstein Irreducibility Criterion is described in Cox (2011).

Chapter 10
Transcendence of e and π

The purpose of this chapter is to prove that the number π is transcendental, thereby completing the proof of the impossibility of squaring the circle; that is Problem III of the Introduction. We first prove that e is a transcendental number, which is somewhat easier. This is of considerable interest in its own right, and its proof introduces many of the ideas which will be used in the proof for π. With the aid of the theory of symmetric polynomials we can then modify the proof for e to give the proof for π.

The proof for e uses elementary facts about integration. (We expect you are familiar with these facts from a course on calculus.) The proof for π uses integration of complex-valued functions, which we introduce in this chapter.

10.1 Preliminaries

This section contains the preliminary results which are needed for our proofs that e and π are transcendental numbers. The extra preliminaries which are used only in the proof for π will be given later.

Factors and Prime Numbers

Let a and b be integers. Recall that b is said to be a *factor* of a if there is an integer c such that $a = bc$.

The integer 0 is exceptional in that it has every integer as a factor (if $a = 0$ then every b satisfies $a = bc$ for $c = 0$). Another way to say this is that

if a is an integer which fails to have some integer b as a factor, then $a \neq 0$.

This statement provides a way of proving that an integer is not zero, and it will play an important rôle in our proof. This leads us to consider ways in which we can show that a given integer fails to have some other given integer as a factor.

S. A. Morris et al., *Abstract Algebra and Famous Impossibilities*, Undergraduate Texts in Mathematics, https://doi.org/10.1007/978-3-031-05698-7_10

One consequence of the definition of a factor is that if a is nonzero and has b as a factor then $|a| \geq |b|$. Hence

if $|b| > |a|$ *and* $a \neq 0$, *then* b *cannot be a factor of* a.

Another consequence of the definition of a factor is that if an integer b is a factor of two integers, then it is also a factor of their sum, their difference and their product. Hence

if b *is a factor of an integer* a_1 *but* b *is not a factor of an integer* a_2, *then* b *is not a factor of* $a_1 + a_2$

as otherwise b would be a factor of the difference $(a_1 + a_2) - a_1 = a_2$.

Recall that an integer p is said to be a *prime* if $p > 1$ and if p has no positive factors other than 1 and p itself. The Fundamental Theorem of Arithmetic is stated in Exercises 1.4 #9. It can be shown from this theorem that if a prime number p is a factor of the product ab of two integers a and b, then p must be a factor of a or a factor of b (or a factor of both a and b).

From this it is easy to deduce another useful test for failure to be a factor of a product:

if a prime p *is not a factor of any of the integers* $a_1, a_2, \ldots, a_m$ *then it is not a factor of their product* $a_1 a_2 \ldots a_m$.

Another fact which we shall need later is given by the following proposition. A proof of this result was known to the Greek mathematician Euclid about 2,400 years ago. For alternative proofs see Aigner and Ziegler (2018).

10.1.1 Proposition. *There are infinitely many prime numbers.*

Proof. To prove this we note that there is at least one prime number (2 – for example) and we shall prove that given any finite set of prime numbers

$$\{p_1, p_2, \ldots, p_n\} \tag{1}$$

there must be another prime number which is not in the set. To this end consider the integer

$$K = p_1 p_2 \ldots p_n + 1. \tag{2}$$

Each p_i is a factor of the first term on the right hand side of (2), but not of the second term, 1. Hence p_i is not a factor of K. But K does have a prime factor q by the Fundamental Theorem of Arithmetic. Thus there is a prime q which is not in the set (1), as we wished to show. ■

This means that, given any integer m, no matter how large, we can be sure that there is a prime (indeed, infinitely many primes) larger than m. We use this fact frequently in this chapter.

Calculus

Many readers will have met the following properties of integrals in a calculus course. (Proofs can be found in many texts: see for example Spivak (1967, Theorems 5 and 6 in Chapter 13).)

10.1.2 Theorem. *Let $a, b \in \mathbb{R}$ with $a < b$ and let f, g be functions from the interval $[a, b]$ into $\mathbb{R}$.*

(i) [Linearity of Integration] *If $d_1, d_2 \in \mathbb{R}$ and if each of the following integrals exists, then*

$$\int_a^b (d_1 f(x) + d_2 g(x))\,dx = d_1 \int_a^b f(x)dx + d_2 \int_a^b g(x)dx.$$

(ii) [Integration by Parts] *If f and g are functions such that each of the following integrals exists, then*

$$\int_a^b f(x)g'(x)dx = f(b)g(b) - f(a)g(a) - \int_a^b f'(x)g(x)dx. \qquad \blacksquare$$

We shall use these results only in the cases where the functions are polynomial or exponential functions, or products thereof; the integrals of such functions exist.

The proof of the next result depends on the idea of the degree of a nonzero polynomial form, which was explained in Chapter 1.

10.1.3 Lemma. *Let $g(x)$ and $h(x)$ be polynomials whose coefficients are real numbers. If*

$$g(x) = h(x)\mathrm{e}^{-x}, \quad for\ all\ x \in \mathbb{R}, \tag{3}$$

then $g(x)$ and $h(x)$ are both equal to the zero polynomial.

Proof. It follows from (3) that $h(x) = g(x)\mathrm{e}^x$ and so differentiating both sides gives

$$h'(x) = \big(g(x) + g'(x)\big)\,\mathrm{e}^x, \quad \text{for all } x \in \mathbb{R}.$$

If we multiply both sides by $g(x)$ and then replace the factor $\mathrm{e}^x g(x)$ on the right hand side by $h(x)$, we find that

$$\begin{aligned} h'(x)g(x) &= \big(g(x) + g'(x)\big)\, e^x g(x) \\ &= g(x)h(x) + g'(x)h(x), \qquad \text{for all } x \in \mathbb{R}. \end{aligned}$$

Rearranging this equation and interpreting it in terms of polynomial forms gives

$$g(X)h(X) = h'(X)g(X) - g'(X)h(X). \tag{4}$$

If either of the polynomials $g(X)$ or $h(X)$ is zero, then so is the other by (3); hence it only remains to prove the lemma in the case where both $g(X)$ and $h(X)$ are nonzero polynomials.

Now the degree of $g'(X)$ is either undefined or is one less than the degree of $g(X)$, and similarly the degree of $h'(X)$ is either undefined or is one less than the degree of $h(X)$. Hence the equation (4) gives a contradiction because the degree of its right side either is undefined or is less than the degree of $g(X)h(X)$. Thus the case in which both of the polynomials $g(X)$ and $h(X)$ are nonzero cannot occur. ■

The following lemma concerning the limit of a sequence will also be needed.

10.1.4 Lemma. *For each real number* $c \geq 0$, $\lim_{n\to\infty} \frac{c^n}{n!} = 0$.

Proof. Let m be an integer such that $m > 1$ and $m \geq 2c$. For all integers $n \geq m$,

$$\begin{aligned}\frac{c^n}{n!} &= \frac{c}{1}\frac{c}{2}\cdots\frac{c}{m-1}\frac{c}{m}\frac{c}{m+1}\cdots\frac{c}{n} \\ &\leq \frac{c}{1}\frac{c}{2}\cdots\frac{c}{m-1}\frac{1}{2}\;\frac{1}{2}\cdots\frac{1}{2} = \frac{d}{2^{n-m+1}}\end{aligned}$$

where d is the constant $\frac{c}{1}\frac{c}{2}\cdots\frac{c}{m-1}$.

Since $\frac{d}{2^{n-m+1}}$ approaches 0 as n approaches ∞, so does $\frac{c^n}{n!}$. ■

Expanding a polynomial in powers of $(X - r)$

Let $g(X)$ be a polynomial over a field so that

$$g(X) = c_0 + c_1X + c_2X^2 + \ldots + c_mX^m \tag{5}$$

for some integer $m \geq 0$ and coefficients $c_0, c_1, c_2, \ldots, c_m$ in the field. It is often very useful to know that, given r in the field, we can express $g(X)$ in the form

$$g(X) = b_0 + b_1(X - r) + b_2(X - r)^2 + \ldots + b_m(X - r)^m \tag{6}$$

where the coefficients $b_0, b_1, b_2, \ldots, b_m$ are also in the field.

Whereas (5) expresses the polynomial as a linear combination of powers of X, (6) expresses it as a linear combination of the powers of $(X - r)$ and is called the *expansion of* $g(X)$ *in powers of* $(X - r)$.

A general procedure for expanding a polynomial in powers of $(X - r)$ is to expand each monomial X^i in powers of $(X - r)$ using the Binomial Theorem, and then collect like terms. We illustrate this procedure in the next example.

Note that the coefficients b_i in (6) turn out to be polynomials $d_i(X)$ evaluated at r; that is, $b_i = d_i(r)$.

10.1.5 Example. To expand the polynomial (over $\mathbb{R}$)

$$g(X) = 3 + 4X + 7X^2 + X^3$$

in powers of $(X - r)$, we write

$$\begin{aligned} g(X) &= g(X - r + r) \\ &= 3 + 4[(X - r) + r] + 7[(X - r) + r]^2 + [(X - r) + r]^3 \\ &= 3 + 4[(X - r) + r] + 7[(X - r)^2 + 2r(X - r) + r^2] + \\ &\qquad [(X - r)^3 + 3r(X - r)^2 + 3r^2(X - r) + r^3] \\ &= (3 + 4r + 7r^2 + r^3) + (4 + 14r + 3r^2)(X - r) + \\ &\qquad (7 + 3r)(X - r)^2 + (X - r)^3. \end{aligned}$$

Thus we have shown that the polynomial $g(X)$ can be expanded in powers of $(X - r)$ to give

$$g(X) = d_0(r) + d_1(r)(X - r) + d_2(r)(X - r)^2 + d_3(r)(X - r)^3$$

where $d_0(r)$, $d_1(r)$, $d_2(r)$, and $d_3(r)$ are the values at r of polynomials $d_0(X), \ldots, d_3(X)$ given by

$$\begin{aligned} d_0(X) &= 3 + 4X + 7X^2 + X^3, \\ d_1(X) &= 4 + 14X + 3X^2, \\ d_2(X) &= 7 + 3X, \\ d_3(X) &= 1. \end{aligned}$$

In particular we have, as in (6),

$$g(X) = b_0 + b_1(X - r) + b_2(X - r)^2 + b_3(X - r)^3$$

where $b_0 = d_0(r)$, $b_1 = d_1(r)$, $b_2 = d_2(r)$, and $b_3 = d_3(r)$. ■

The general result, whose proof is obvious from the above example, is as follows.

10.1.6 Proposition. *If $g(X)$ is a polynomial of degree at most n, whose coefficients are integers, then there exist polynomials*

$$d_0(X), \ldots, d_n(X)$$

with integer coefficients and degree at most n such that, for all $r \in \mathbb{R}$,

$$g(X) = d_0(r) + d_1(r)(X - r) + \ldots + d_n(r)(X - r)^n.$$ ■

Note that the same polynomials $d_0(X), \ldots, d_n(X)$ work for all values of r.

The following result confirms that when two polynomials in $(X - r)$ are equal, we can equate coefficients (just as we can for ordinary polynomials, which is the case $r = 0$).

10.1.7 Lemma. *Let $r \in \mathbb{R}$ and $b_i, c_i \in \mathbb{R}$ for $i \in \{0, 1, \ldots, n\}$. If $c_0 + c_1(X - r) + \ldots + c_n(X - r)^n = b_0 + b_1(X - r) + \ldots + b_n(X - r)^n$, then*

$$c_0 = b_0,\ c_1 = b_1, \ldots, c_n = b_n.$$

Proof. We have

$$c_0 + c_1(x - r) + \ldots + c_n(x - r)^n = b_0 + b_1(x - r) + \ldots + b_n(x - r)^n$$

for all $x \in \mathbb{R}$. Substituting $x = r$ gives $c_0 = b_0$. Differentiating with respect to x and substituting $x = r$ then gives $c_1 = b_1$. Repeating this a further $n - 1$ times gives the required result. ∎

Approximating Real Numbers by Rational Numbers

That we can approximate any real number a by rational numbers is well-known. We say that the set of rational numbers $\mathbb{Q}$ is a *dense* subset of the real line $\mathbb{R}$; that is, given any interval containing a, there is a rational number which lies in that interval. More formally this means that, given any real number a and a "small" number $\varepsilon > 0$, we can find integers M and M_1 and a number ε_1 such that $|\varepsilon_1| < \varepsilon$ and

$$a = \frac{M_1}{M} + \varepsilon_1. \tag{7}$$

This says that there is a rational number M_1/M at a distance of $|\varepsilon_1| < \varepsilon$ from the given real number a.

For our purposes, a more relevant type of approximation is suggested by the following question: *Given a "small" number ε, can we find integers M and M_1 and a nonzero number ε_1 with $|\varepsilon_1| < \varepsilon$ such that*

$$a = \frac{M_1 + \varepsilon_1}{M}? \tag{8}$$

Thus we are asking not only that the distance from a to M_1/M be small, but also that it be small even when multiplied by M, the denominator of the approximating fraction.

10.1.8 Example. It is easy to see that the number e can be approximated as in (8). To see this first note that the well-known Taylor series for the exponential function gives

$$\mathrm{e} = 1 + \frac{1}{1!} + \frac{1}{2!} + \ldots + \frac{1}{n!} + \frac{1}{(n+1)!} + \ldots.$$

For each integer $n \geq 1$, let

$$A_n = 1 + \frac{1}{1!} + \frac{1}{2!} + \ldots + \frac{1}{n!}.$$

Hence

$$\begin{aligned} \mathrm{e} &= A_n + \frac{1}{(n+1)!}\left[1 + \frac{1}{n+2} + \frac{1}{(n+2)(n+3)} + \ldots\right] \\ &\leq A_n + \frac{1}{(n+1)!}\left[1 + \frac{1}{2} + \frac{1}{2^2} + \ldots\right] \\ &= A_n + \frac{2}{(n+1)!}. \end{aligned}$$

Given $\varepsilon > 0$ choose n to be any integer $> 2/\varepsilon$, so that $2/n < \varepsilon$. Put $M = n!$ and $M_1 = (n!)A_n$. You should now be able to see how to choose ε_1 so that (8) holds. We leave the details as Exercises 10.1 #8. ■

Can all real numbers a be approximated as in (8) by rationals? That the answer to this question is NO is clear from the following proposition.

10.1.9 Proposition. *Let a be a real number with the following property: for each real number $\varepsilon > 0$ there are integers M and M_1, and a nonzero real number ε_1 with $|\varepsilon_1| < \varepsilon$ such that*

$$a = \frac{M_1 + \varepsilon_1}{M}.$$

Then the real number a is irrational.

Proof. Suppose, contrary to the conclusion of the proposition, that a is a rational number, say $a = p/q$ where p and q are integers and $q > 0$. Let $\varepsilon = 1/(2q)$. By the hypothesis of the proposition there exist integers M and M_1, and a nonzero real number ε_1 with $|\varepsilon_1| < \varepsilon$ such that

$$\frac{p}{q} = \frac{M_1 + \varepsilon_1}{M}, \quad \text{and hence} \quad pM - qM_1 = q\varepsilon_1.$$

The left hand side of the last equation is an integer, while $0 < |q\varepsilon_1| < \frac{1}{2}$. This contradiction shows that a cannot be rational. ■

From Example 10.1.8 and Proposition 10.1.9 it follows that e is an irrational number. Note also that requiring ε_1 to be nonzero is crucial in the proof of the above proposition. By requiring ε_1 to be nonzero, we are not allowing the rational approximation M_1/M to equal the number a.

A very useful idea which this proposition suggests is that
the "better" a real number can be approximated by rational numbers (not equal to it), the "worse" that real number must be.

Here we are thinking of the rational numbers as "good", the irrational numbers which are algebraic as "bad", and the irrational numbers which are not algebraic as "worse". We shall follow up this idea in the next section.

Exercises 10.1

1. Let n be a positive integer. Verify that each of the following numbers has $(n-1)!$ as a factor.

   (i) $n!$.
   (ii) $n! + (n+1)! + \ldots + (n+m)!$, where m is a positive integer.
   (iii) $(2n)!$.

2. Prove, from the definition of factor, that if an integer b is a factor of another integer a which is not zero, then $|a| \geq |b|$.

3. Prove, from the definition of factor, that if an integer b is a factor of each of the integers a_1 and a_2 then b is also a factor of the integers $a_1 + a_2$, $a_1 - a_2$, and $a_1 a_2$. Is b a factor of a_1/a_2 in every case?

4. Let m and n be integers with $m > n$. In each of the following cases state whether the integer must have m as a factor. (Justify your answers.)

   (i) $mn - n$.
   (ii) $nm - m$.

5. Let p be a prime number and let m be a positive integer such that $m < p$. Show that p is not a factor of

   (i) $m(m-1)$;
   (ii) $m!$;
   (iii) $(m!)^p$.

6. Let $f(x) = a_0 + a_1 x + \ldots + a_n x^n$ be a polynomial function with real coefficients. The result of differentiating this polynomial i times in succession is denoted by $f^{(i)}$. It is called the *i-th derivative* of the polynomial f. Show that $f^{(n+1)}(x) = 0$ for all $x \in \mathbb{R}$.

7. (a) Express $f(X) = 2 - 4X + 3X^2 - 7X^3$ as a polynomial in $(X - r)$. Write down the formulae for the polynomials $d_0(X), \ldots, d_3(X)$ in Proposition 10.1.6.
   (b) Repeat (a) with $f(X) = c_0 + c_1 X + c_2 X^2 + c_3 X^3 + c_4 X^4$ where $c_0, \ldots, c_4 \in \mathbb{Z}$.

8. Complete the details of Example 10.1.8. Why is M_1 an integer?

9. (a) Show that the rational number 3/7 can be approximated as in (7) of Section 10.1.
   [Hint. What can M_1 and M be if $\varepsilon = 1/100$?]
   (b) Show that every rational number can be approximated as in (7) of Section 10.1.

10. Would Proposition 10.1.9 be true if the word "nonzero" were omitted? (Justify your answer.)
11. Prove that each of the numbers cos(1) and sin(1) is irrational by using the Taylor series for the functions cos and sin.

10.2 e is Transcendental

An important property of the number e is that, if $g(x) = \mathrm{e}^x$ for all $x \in \mathbb{R}$, then $g'(x) = g(x)$ for all x. Indeed this fact distinguishes the number e from all other positive real numbers (as shown in Exercises 10.2 #1 below). We shall also use the fact that $\mathrm{e}^{x+y} = \mathrm{e}^x \mathrm{e}^y$ for all real numbers x and y (a property of e which is shared by all positive reals).

Our proof that the number e is transcendental (see Definition 2.1.5) shall begin by supposing on the contrary that e is algebraic over $\mathbb{Q}$. We let

$$t(X) = a_0 + a_1 X + \ldots + a_{m-1} X^{m-1} + X^m$$

denote $\mathrm{irr}(\mathrm{e}, \mathbb{Q})$, the irreducible polynomial of e over $\mathbb{Q}$. Then we have $m \geq 1$ and $a_i \in \mathbb{Q}$ for each i, and

$$0 = t(\mathrm{e}) = a_0 + a_1 \mathrm{e} + \ldots + a_{m-1} \mathrm{e}^{m-1} + \mathrm{e}^m. \tag{*}$$

Now clearly $a_0 \neq 0$ as otherwise X would be a factor of the polynomial $t(X)$, which is irreducible (and which is not just X since $\mathrm{e} \neq 0$).

If we now multiply $(*)$ by the product of the denominators of the rational numbers $a_0, a_1, \ldots, a_{m-1}$ we get

$$c_0 + c_1 \mathrm{e} + \ldots + c_m \mathrm{e}^m = 0 \tag{1}$$

where $c_0, c_1, \ldots, c_m$ are integers such that $c_0 \neq 0$ and $c_m \neq 0$.

In due course, we shall derive a contradiction from this.

Idea of Proof: M's and ε's

To show that (1) leads to a contradiction, we shall use the idea (mentioned in Section 10.1) that the "worse" a number is, the "better" it can be approximated by rationals. As we have seen in Example 10.1.8 and Proposition 10.1.9, the irrationality of e follows very simply from the following property: *For each $\varepsilon > 0$ there is a nonzero number ε_1 with $|\varepsilon_1| < \varepsilon$ and integers M and M_1 such that*

$$\mathrm{e} = \frac{M_1 + \varepsilon_1}{M}.$$

The aim now is to extend this "close approximation" by rationals to each of the remaining powers of e in the equation (1). The aim is thus:
Given $\varepsilon > 0$, *to show there are numbers* $\varepsilon_1, \ldots, \varepsilon_m$ *all less than* ε *in absolute value, and integers* $M, M_1, \ldots, M_m$ *such that*

$$\mathrm{e} = \frac{M_1 + \varepsilon_1}{M}, \quad \mathrm{e}^2 = \frac{M_2 + \varepsilon_2}{M}, \quad \ldots \quad , \quad \mathrm{e}^m = \frac{M_m + \varepsilon_m}{M}.$$

Substituting these powers of e into (1) and rearranging the terms gives

$$[c_0 M + c_1 M_1 + \ldots + c_m M_m] + [c_1\varepsilon_1 + \ldots + c_m\varepsilon_m] = 0. \tag{2}$$

Because M is an integer and so is each c_i and each M_i, the first sum on the left side of (2) is an integer. If we can show this integer is nonzero we shall get the desired contradiction: we can choose each ε_i so small that the second sum on the left side of (2) has its absolute value less than $\frac{1}{2}$ and hence is unable to cancel out the first sum.

It now remains to show how to find suitable numbers $\varepsilon_1, \ldots, \varepsilon_m$ and $M, M_1, \ldots, M_m$.

Producing the ε's and the M's from integrals

The following lemma shows how an integer (namely $k!$) arises from an integral involving the exponential function. This lemma therefore suggests the possibility of using integrals to construct the required integers $M, M_1, \ldots, M_m$.

10.2.1 Lemma. *For each integer* $k \geq 0$ *there are polynomials* $g_k(X)$ *and* $h_k(X)$ *with real coefficients such that, for all* $r \in \mathbb{R}$,

$$(a) \qquad \int_0^r x^k \mathrm{e}^{-x} dx = k! - \mathrm{e}^{-r} g_k(r),$$

$$(b) \qquad \int_0^r (x-r)^k \mathrm{e}^{-x} dx = h_k(r) - \mathrm{e}^{-r} k!.$$

Proof. This can be proved by mathematical induction on k, using integration by parts (see Theorem 10.1.2). ■

Motivated by the above lemma, we shall aim at getting the integers M and $M_1, \ldots, M_m$ from integrals of the form

$$\int_0^r f(x)\mathrm{e}^{-x}\, dx \tag{3}$$

for $r \in \{1, 2, \ldots, m\}$, where f is a polynomial function. In the following example, we show how to apply the above lemma to such integrals.

10.2.2 Example. Let $f(X) = 3 + 4X - 10X^3$. By linearity of integration (Theorem 10.1.2), and then Lemma 10.2.1(a), for all $r \in \mathbb{R}$,

$$
\begin{aligned}
\int_0^r & f(x)\mathrm{e}^{-x}\,dx \\
&= \int_0^r (3+4x-10x^3)\mathrm{e}^{-x}dx \\
&= 3\int_0^r 1.\mathrm{e}^{-x}dx + 4\int_0^r x\mathrm{e}^{-x}dx - 10\int_0^r x^3\mathrm{e}^{-x}dx \\
&= 3(0! - \mathrm{e}^{-r}g_0(r)) + 4(1! - \mathrm{e}^{-r}g_1(r)) - 10(3! - \mathrm{e}^{-r}g_3(r)) \\
&= (3.0! + 4.1! - 10.3!) - \mathrm{e}^{-r}(3g_0(r) + 4g_1(r) - 10g_3(r))
\end{aligned}
$$

where $g_0(X)$, $g_1(X)$ and $g_3(X)$ are polynomials. Hence

$$\int_0^r f(x)\mathrm{e}^{-x}dx = M - \mathrm{e}^{-r}G(r)$$

where $M = 3.0! + 4.1! - 10.3! = -53$ and $G(X)$ is the polynomial

$$G(X) = 3g_0(X) + 4g_1(X) - 10g_3(X).$$

■

The next lemma generalizes this example.

10.2.3 Lemma. *Given a polynomial $f(X)$ with real coefficients, there is a unique real number M and a unique polynomial $G(X)$ with real coefficients such that, for all $r \in \mathbb{R}$,*

$$\int_0^r f(x)e^{-x}dx = M - e^{-r}G(r).$$

Indeed M and $G(X)$ are unique in the strong sense that, if $P_1(X)$ and $P_2(X)$ are polynomials with real coefficients such that

$$\int_0^r f(x)e^{-x}dx = P_1(r) - \mathrm{e}^{-r}P_2(r)$$

for all $r \in \mathbb{R}$, then $P_2(X) = G(X)$ and $P_1(X) = M$ (so that $P_1(X)$ is a constant polynomial).

Proof. The existence of M and $G(X)$ follows as in Example 10.2.2 by expanding $f(X)$ in powers of X, using linearity of integration (Theorem 10.1.2(i)) and then applying Lemma 10.2.1(a).

To prove uniqueness, we assume that the two formulae in the statement of the lemma hold. Then, for all $r \in \mathbb{R}$,

$$M - \mathrm{e}^{-r}G(r) = P_1(r) - \mathrm{e}^{-r}P_2(r)$$

and hence $M - P_1(r) = \mathrm{e}^{-r}(G(r) - P_2(r))$. It follows from Lemma 10.1.3 that $M - P_1(X)$ and $G(X) - P_2(X)$ must both be the zero polynomial, which gives $P_1(X) = M$ and $P_2(X) = G(X)$, as required. ■

We now use this lemma to define the numbers we need for the transcendence proof.

10.2.4 Definition. In Lemma 10.2.3, choose f to be the polynomial

$$f(X) = \frac{X^{p-1}}{(p-1)!}(X-1)^p(X-2)^p \ldots (X-m)^p \tag{4}$$

where p is a prime number (which we shall later take to be rather large) and m is the integer occurring in (1). Further, choose

(i) M and $G(X)$ to be the number and polynomial given by Lemma 10.2.3,
(ii) $M_r = G(r)$ for $r \in \{1, 2, \ldots, m\}$, and
(iii) $\varepsilon_r = e^r \int_0^r f(x)e^{-x}dx$ for $r \in \{1, 2, \ldots, m\}$. ■

Note that, from (4), the degree n of $f(X)$ is

$$n = mp + p - 1. \tag{5}$$

Also, by Proposition 10.1.6, there are polynomials $d_0, \ldots, d_n$ with integer coefficients and degree at most n such that, for all $r \in \mathbb{R}$,

$$f(X) = \frac{1}{(p-1)!}\left(d_0(r) + d_1(r)(X-r) + \ldots + d_n(r)(X-r)^n\right). \tag{6}$$

10.2.5 Lemma. *With the notation as in (6), if* $r \in \{1, 2, \ldots, m\}$ *then* $d_0(r) = d_1(r) = \ldots = d_{p-1}(r) = 0$.

Proof. Fix $r \in \{1, 2, \ldots, m\}$. Because $(X-r)^p$ is a factor of $f(X)$ by (4), we have

$$f(X) = \frac{1}{(p-1)!}(X-r)^p g(X)$$

for some polynomial $g(X)$ with integer coefficients. If we expand $g(X)$ in powers of $(X-r)$ (see Proposition 10.1.6) and multiply each resulting term by $(X-r)^p$ we see that

$$f(X) = \frac{1}{(p-1)!}\left(b_p(X-r)^p + b_{p+1}(X-r)^{p+1} + \ldots + b_n(X-r)^n\right) \tag{7}$$

for some real numbers $b_p, \ldots, b_n$. By Lemma 10.1.7 we can equate the coefficients of 1, $(X-r), \ldots, (X-r)^n$ in (7) and (6). Since the coefficients of 1, $(X-r), \ldots, (X-r)^{p-1}$ are zero in (7), they must be zero in (6) also. ■

Properties of $M, M_1, \ldots, M_m$ and $\varepsilon_1, \ldots, \varepsilon_m$

We now derive properties of the numbers which were introduced in Definition 10.2.4.

10.2.6 Lemma. *(a) The number M is an integer which does not have the prime p as a factor when $p > m$.*

(b) There is a polynomial G_1 with integer coefficients and degree at most n such that $G(r) = pG_1(r)$ for $r \in \{1, 2, \ldots, m\}$. In particular, $M_1, \ldots, M_m$ are all integers with the prime p as a factor.

Proof. (a) The polynomial $f(X)$ defined in (4) can be expanded in powers of X so that we can write

$$f(X) = \frac{1}{(p-1)!}\left(a_{p-1}X^{p-1} + a_pX^p + \ldots + a_nX^n\right) \tag{8}$$

where $a_{p-1}, \ldots, a_n$ are integers with $a_{p-1} = \pm(m!)^p \neq 0$.

By the linearity of integration (Theorem 10.1.2(i)), it follows from (8) that

$$\begin{aligned}
&\int_0^r f(x)\mathrm{e}^{-x}\,dx \\
&= \int_0^r \frac{1}{(p-1)!}\left(a_{p-1}x^{p-1} + a_px^p + \ldots + a_nx^n\right)\mathrm{e}^{-x}dx \\
&= \frac{1}{(p-1)!}\left(a_{p-1}\int_0^r x^{p-1}\mathrm{e}^{-x}dx + a_p\int_0^r x^p\mathrm{e}^{-x}dx + \ldots + a_n\int_0^r x^n\mathrm{e}^{-x}dx\right).
\end{aligned}$$

By Lemma 10.2.1(a), each of the integrals in the above sum is the difference of two terms, the second of which involves e^{-r}. Hence, by the method used in Example 10.2.2, $\int_0^r f(x)\mathrm{e}^{-x}dx$ is the difference of two sums, the first of which is

$$\frac{1}{(p-1)!}\left(a_{p-1}(p-1)! + a_pp! + \ldots + a_nn!\right)$$

while the second is of the form $\mathrm{e}^{-r} \times$ *polynomial in* r. By (i) of Definition 10.2.4 and the uniqueness part of Lemma 10.2.3, M is this first term. Hence, dividing $(p-1)!$ into each term in the above sum gives

$$M = a_{p-1} + a_pp + \ldots + a_nn(n-1)\ldots(p+1)p.$$

Since the coefficients $a_{p-1}, a_p, \ldots, a_n$ are all integers, the number M is an integer. Also the prime p is a factor of every term in the above sum except the first term $a_{p-1} = \pm(m!)^p$, which cannot contain p as a factor when $p > m$ (by Exercises 10.1 #5); hence p cannot be a factor of M when $p > m$.

(b) [Here we use the expansion of $f(X)$ in powers of $X - r$, as given in (6) above.] By the linearity of integration, it follows from (6) that

$$\int_0^r f(x)e^{-x}dx = \frac{1}{(p-1)!}\left(d_0(r)\int_0^r e^{-x}dx + d_1(r)\int_0^r (x-r)e^{-x}dx + \ldots + d_n(r)\int_0^r (x-r)^n e^{-x}dx\right).$$

By Lemma 10.2.1(b), each of the integrals in the above sum is the difference of two terms, the second of which involves e^{-r}. Hence $\int_0^r f(x)\mathrm{e}^{-x}dx$ is the difference of two sums. The second of these is

$$\frac{\mathrm{e}^{-r}}{(p-1)!}(d_0(r) + d_1(r)1! + \ldots + d_n(r)n!) \tag{9}$$

and the first is a polynomial in r with real coefficients. By the uniqueness part of Lemma 10.2.3, $G(r)$ is equal to the polynomial part of (9); that is,

$$G(r) = \frac{1}{(p-1)!}(d_0(r) + d_1(r)1! + \ldots + d_n(r)n!)\,.$$

In the special cases where $r \in \{1, 2, \ldots, m\}$, we know from Lemma 10.2.5 that $d_0(r) = d_1(r) = \ldots = d_{p-1}(r) = 0$. Hence

$$G(r) = \frac{1}{(p-1)!}\left(d_p(r)p! + \ldots + d_n(r)n!\right).$$

If we divide $(p-1)!$ into each term we see that $G(r) = pG_1(r)$ where

$$G_1(X) = d_p(X) + d_{p+1}(X)(p+1) + \ldots + d_n(X)n(n-1)\ldots(p+1).$$

Notice that $G_1(X)$ is a polynomial with integer coefficients and degree at most n since each $d_i(X)$ is such a polynomial. Hence $G_1(r)$ is an integer (since r is) and so, by (ii) of Definition 10.2.4, $M_r = pG_1(r)$ is an integer having p as a factor. ■

The next lemma shows that the ε's defined in Definition 10.2.4 can be made arbitrarily small by choosing p large enough.

10.2.7 Lemma. *If $f(X)$ is given as a function of p by (4), and $r \in \{1, 2, \ldots, m\}$, then*

$$\lim_{p\to\infty}\int_0^r f(x)\mathrm{e}^{-x}\,dx = 0.$$

Proof. Let $r \in \{1, 2, \ldots, m\}$ and let $x \in [0, r]$. Using (4) gives

$$\begin{aligned}|f(x)\mathrm{e}^{-x}| &\le |f(x)|\\ &= \frac{x^{p-1}}{(p-1)!}\left|(x-1)^p(x-2)^p\ldots(x-m)^p\right|\end{aligned}$$

$$\leq \frac{x^{p-1}}{(p-1)!}(x+1)^p(x+2)^p \dots (x+m)^p$$
$$\leq \frac{r^{p-1}}{(p-1)!}(r+1)^p(r+2)^p \dots (r+m)^p$$
$$= \frac{1}{r}\frac{C^p}{(p-1)!},$$

where $C = r(r+1)(r+2)\dots(r+m)$ is independent of p. But

$$\left|\int_0^r f(x)\mathrm{e}^{-x}\,dx\right| \leq (length\ of\ interval) \times (upper\ bound\ for\ |f(x)\mathrm{e}^{-x}|)$$
$$= r \times \frac{1}{r}\frac{C^p}{(p-1)!}.$$

Hence the conclusion of the lemma follows from Lemma 10.1.4. ■

With the aid of the above lemmas, it is now an easy task to prove that e is transcendental.

10.2.8 Theorem. e *is a transcendental number.*

Proof. Suppose on the contrary that e is algebraic. Then, as shown at the start of this section (see (1) at the beginning of this section), there is an integer $m \geq 1$ and integers $c_0, c_1, c_2, \dots, c_m$ such that $c_0 \neq 0$, $c_m \neq 0$ and

$$c_0 + c_1\mathrm{e} + c_2\mathrm{e}^2 + \dots + c_m\mathrm{e}^m = 0. \tag{10}$$

Let M and M_r, ε_r, for $r \in \{1, 2, \dots, m\}$, be as in Definition 10.2.4. From parts (iii), (i) and (ii) of that definition respectively, it follows that

$$\mathrm{e}^{-r}\varepsilon_r = \int_0^r f(x)\mathrm{e}^{-x}dx = M - \mathrm{e}^{-r}G(r) = -M - \mathrm{e}^{-r}M_r$$

and hence

$$\mathrm{e}^r = \frac{M_r + \varepsilon_r}{M}.$$

Substituting these results into (10) gives

$$[c_0M + c_1M_1 + \dots + c_mM_m] + [c_1\varepsilon_1 + \dots + c_n\varepsilon_m] = 0. \tag{11}$$

We now choose the prime number p to be larger than each of m and $|c_0|$. As $p > m$ it follows from Lemma 10.2.6(a) that

the integer M does not have p as a factor.

As $p > |c_0|$ it follows that c_0 does not have p as a factor and hence (as explained at the beginning of Section 7.1) that

$$c_0 M \textit{ does not have } p \textit{ as a factor.}$$

But each of $M_1, M_2, \ldots, M_m$, is an integer having p as a factor by Lemma 10.2.6(b). Hence p is not a factor of the sum

$$c_0 M + c_1 M_1 + \ldots + c_m M_m, \textit{ which is therefore a nonzero integer.} \tag{12}$$

On the other hand, if we choose p sufficiently large (which is possible by Proposition 10.1.1), then by Definition 10.2.4(iii) and Lemma 10.2.7 we get

$$|c_1\varepsilon_1 + \ldots + c_m\varepsilon_m| < \frac{1}{2}. \tag{13}$$

But (12) and (13) contradict (11). So our supposition is wrong. Hence e is a transcendental number. ■

10.2.9 Remark. Now that the proof is complete, you might like to reflect on the reasons for the various features in the definition of $f(X)$ in (4).

(i) The $(p-1)!$ in the denominator means that ε_r can be made arbitrarily small by taking p large enough. (See Lemma 10.2.7.)
(ii) The factor X^{p-1} ensures that, despite the $(p-1)!$ in the denominator of $f(X)$, M is an integer. The exact power $p-1$ (rather than a higher power) means that M is not divisible by p if $p > m$. (See the proof of Lemma 10.2.6(a). See also Exercise 3 below.)
(iii) The factors $(X-1)^p, \ldots, (X-m)^p$ ensure that, despite the $(p-1)!$ in the denominator, each of $M_1, \ldots, M_m$ is an integer. (See the proof of Lemma 10.2.6(b).)
(iv) All of the factors $(X-1)^p, \ldots, (X-m)^p$ are required so that we can deal with all the exponents $1, 2, \ldots, m$ in (10).

Exercises 10.2

1. [This exercise proves the claim made at the beginning of Section 10.2 that e is distinguished from all other positive real numbers by the fact that the derivative of e^x is e^x.]
   Let c be a positive real number and assume that $g(x) = c^x$ for all $x \in \mathbb{R}$. If $g'(x) = g(x)$ for all real x, prove that $c = e$.
   [Hint. $c^x = (e^{\ln c})^x = e^{x \ln c}$.]
2. Prove Lemma 10.2.1.
3. Show that Lemma 10.2.6 would be false if we omitted the factor X^{p-1} in the definition of the function $f(X)$ given in (4) of the text.
4. Show that Lemma 10.2.6 would be false if we replaced the factor X^{p-1} by the factor X^p in the definition of $f(X)$ given in (4) of the text.

5. (a) Prove that e^2 is a transcendental number.
   [Hint. Use Theorem 10.2.8 and the definition of algebraic.]
   (b) Prove that $e^2 + 3e + 1$ is a nonzero transcendental number.
   (c) Prove that if $f(X)$ is a nonzero polynomial with rational coefficients, then $f(e)$ is a nonzero transcendental number.
6. Prove that $\sqrt{e}$ is transcendental.
   [Hint. Use Corollary 4.4.4.]
7. Deduce from Lemma 10.2.1(a) that $\int_0^\infty x^k e^{-x} dx = k!$.
   [Hint. You may assume that $\lim_{r\to\infty} e^{-r} g(r) = 0$ for each polynomial $g(X)$ with real coefficients.]
8. (a) Deduce from Definition 10.2.4, Lemma 10.2.3 and the hint for Exercise 7 that
$$M = \int_0^\infty f(x)e^{-x} dx.$$
   (b) Using (a) deduce from Definition 10.2.4 and Lemma 10.2.3 that
$$M_r = e^r \int_r^\infty f(x)e^{-x} dx.$$
9. Prove by mathematical induction that if $f(X) = a_0 + a_1X + \ldots + a_nX^n$ then
$$\int_0^r f(x)e^{-x} dx = \sum_{i=0}^n a_i i! - e^{-r} \sum_{i=0}^n f^{(i)}(r)$$
   for each $r \in \mathbb{R}$, where $f^{(i)}(r)$ is the i-th derivative of f evaluated at r. [Hint. For each integer m, take the induction statement to be
$$\int_0^r f(x)e^{-x} dx = \sum_{i=0}^m f^{(i)}(0) - e^{-r} \sum_{i=0}^m f^{(i)}(r) - \int_0^r f^{(m)}(x)e^{-x} dx.]$$
10. (a) Prove that e^r is irrational for all $r \in \mathbb{N}$.
    (b) Deduce from (a) and the formula just before (11) in the text that $\varepsilon_1, \ldots, \varepsilon_m$ are nonzero. Why didn't we need to use this fact in the proof of Theorem 10.2.8 (in contrast to the proof of Proposition 10.1.9 where ε_1 being nonzero is essential)?

10.3 π is Transcendental – Part 1

The proof that π is transcendental (to be given in Section 10.4) is similar in many respects to that given in Section 10.2 for e. There is, however, a topic we must introduce before the proof, namely integrals of complex-valued functions. We do so in this section. Then using the theory of symmetric polynomials we introduced in Section 8.3, we can prove π is a transcendental number.

Complex Exponentials

In our proof that π is transcendental, we need to consider e^z where z is a complex number.

10.3.1 Definition. If $z = x + iy$ where $x, y \in \mathbb{R}$, then e^z is defined by

$$\mathrm{e}^z = \mathrm{e}^x(\cos(y) + i\sin(y)).$$

■

10.3.2 Example.

$$\begin{aligned} \mathrm{e}^{1+i} &= \mathrm{e}(\cos(1) + i\sin(1)), \\ \mathrm{e}^{i\pi} &= \mathrm{e}^0(\cos\pi + i\sin\pi) = 1(-1 + 0i) = -1. \end{aligned}$$

This latter fact will be the starting point for our proof that π is transcendental. It is also easy to check that $\mathrm{e}^{z_1}\mathrm{e}^{z_2} = \mathrm{e}^{z_1+z_2}$ for all z_1 and z_2 in $\mathbb{C}$. (Use the expansions of $\cos(y_1 + y_2)$ and $\sin(y_1 + y_2)$.) ■

An equation for π via symmetric polynomials

In the next proposition we need to use the well-known result that every non-constant polynomial with complex coefficients can be written as a product of factors of degree 1. More precisely, if $g(X) \in \mathbb{C}[X]$ has degree n and leading coefficient $c \in \mathbb{C}$ then $g(X)$ can be written as

$$g(X) = c(X - \lambda_1)(X - \lambda_2)\ldots(X - \lambda_n)$$

for complex numbers $\lambda_1, \lambda_2, \ldots, \lambda_n$ which are, of course, the zeros of $g(X)$ in $\mathbb{C}$. This result is known as the Fundamental Theorem of Algebra 8.2.1.

10.3.3 Proposition. *Assume π is algebraic over $\mathbb{Q}$. Then there exist integers $m \geq 1$, $q \geq 1$, $b \geq 1$ and nonzero complex numbers $\beta_1, \beta_2, \ldots, \beta_m$ such that*

$$\mathrm{e}^{\beta_1} + \mathrm{e}^{\beta_2} + \ldots + \mathrm{e}^{\beta_m} + q = 0 \tag{1}$$

and the polynomial

$$h(X) = b(X - \beta_1)(X - \beta_2)\dots(X - \beta_m) \tag{2}$$

has integer coefficients.

Proof. Assume π is algebraic.

Recall that $e^{i\pi} = -1$.

Because this involves $i\pi$ in the exponent rather than just π, we shall consider the irreducible polynomial over $\mathbb{Q}$ of $i\pi$ rather than of π.

Since the complex number i is algebraic over $\mathbb{Q}$ it follows from Corollary 4.4.4 that the complex number $i\pi$ is algebraic (over $\mathbb{Q}$). If $t(X)$ denotes the irreducible polynomial of $i\pi$ over $\mathbb{Q}$ then $t(X)$ is a monic polynomial with rational coefficients such that $t(i\pi) = 0$. The Fundamental Theorem of Algebra 8.2.1 tells us that $t(X)$ factors completely over $\mathbb{C}$ so that

$$t(X) = (X - \alpha_1)(X - \alpha_2)\dots(X - \alpha_k)$$

for some complex numbers $\alpha_1, \alpha_2, \dots, \alpha_k$. Of course one of these α's must be equal to $i\pi$ and we can assume, without loss of generality, that $\alpha_1 = i\pi$.

But $e^{i\pi} = -1$ and hence

$$(e^{\alpha_1} + 1)(e^{\alpha_2} + 1)\dots(e^{\alpha_k} + 1) = 0. \tag{3}$$

You may be puzzled as to why we include the terms $(e^{\alpha_2} + 1), \dots, (e^{\alpha_k} + 1)$ in (3). These are essential because (as you will see later), by including the factors corresponding to all the zeros of f, we are able eventually to obtain symmetric polynomials and then integers.

If we expand the left hand side (LHS) of (3), there are 2^k terms. If we use the fact that $e^{z_1}e^{z_2} = e^{z_1+z_2}$ then we get $2^k - 1$ terms of the form e^{γ} (where γ is a sum of one or more α's) and a single term equal to 1 (the product of all the 1's in (3)).

For example, we see that if $n = 3$ the LHS of (3) is

$$e^{\alpha_1+\alpha_2+\alpha_3} + e^{\alpha_1+\alpha_2} + e^{\alpha_1+\alpha_3} + e^{\alpha_2+\alpha_3} + e^{\alpha_1} + e^{\alpha_2} + e^{\alpha_3} + 1.$$

Thus (3) can be rewritten as

$$e^{\gamma_1} + e^{\gamma_2} + \ldots + e^{\gamma_N} + 1 = 0 \quad (4)$$

(where $N = 2^k - 1$).

Finally, we use the material in Section 8.3 about symmetric polynomials. Symmetric polynomials are the key to our proof that π is a transcendental number.

Now we have met something like the formation of the γ's from the α's before in Proposition 8.3.24. Using that result and its notation,

(i) there is a monic polynomial $t_1(X)$ with rational coefficients which has all the α's as zeros,
(ii) there is a monic polynomial $t_2(X)$ with rational coefficients which has all the sums of two α's as zeros,

and so on, finishing with

(iii) a monic polynomial $t_k(X)$ with rational coefficients which has all the sums of k α's as zeros.

Thus if

$$T(X) = t_1(X)t_2(X)\ldots t_k(X)$$

then $T(X)$ is a monic polynomial with rational coefficients which has all the $N(= 2^k - 1)$ γ's as it zeros. This means that

$$T(X) = (X - \gamma_1)(X - \gamma_2)\ldots(X - \gamma_N). \quad (5)$$

It may happen that some of these γ's (sums of α's) are in fact zero. (For example, it may happen that $\alpha_1 + \alpha_2 + \alpha_3 = 0$). We have no way of knowing if this happens and, if so, how often it happens. But we can allow for it by saying that, if there are q_1 such γ's equal to zero, we can rewrite (4) as

$$e^{\beta_1} + e^{\beta_2} + \ldots + e^{\beta_m} + q = 0$$

where $\beta_1, \beta_2, \ldots, \beta_m$ are all nonzero complex numbers (the γ's which are nonzero) and $q = q_1 + 1$ is a positive integer, which proves (1). Note also that m must be at least 1 since, if $m = 0$, (1) would say that $q = 0$ which is impossible.

From (5)

$$T(X) = X^{q-1}(X - \beta_1)\ldots(X - \beta_m)$$

is a polynomial with rational coefficients and so

$$(X - \beta_1)\dots(X - \beta_m) \tag{6}$$

is also a polynomial with rational coefficients. We can multiply the polynomial in (6) by a suitably large positive integer b (to cancel out all denominators there) to obtain a polynomial

$$\begin{aligned} h(X) &= bX^m + b_{m-1}X^{m-1} + \dots + b_0 \\ &= b(X - \beta_1)(X - \beta_2)\dots(X - \beta_m) \end{aligned}$$

with integer coefficients. (We have omitted the subscript from the leading coefficient b_m to simplify several formulae in Section 10.5.) ■

Exercises 10.3

1. Prove from Definition 10.3.1 that $e^{z_1}e^{z_2} = e^{z_1+z_2}$ for all z_1 and $z_2 \in \mathbb{C}$.
2. If $z = x + iy$ where $x, y \in \mathbb{R}$, show that $|e^z| = e^x$.

10.4 Preliminaries on Complex-valued Integrals

Before we can complete the proof of the transcendence of π (in Section 10.5) we need to introduce integrals of the form

$$\int_a^b f(x)dx$$

where f is a *complex-valued function* (that is, a function $f : \mathbb{R} \to \mathbb{C}$) and a, b are real numbers. First we define integrals of this kind in terms of integrals of real-valued functions (such as you have learned about in your elementary calculus courses). We also define derivatives of complex-valued functions in terms of derivatives of real-valued functions.

10.4.1 Definition. Let a, b be real numbers with $a < b$ and $f : [a, b] \to \mathbb{C}$ a complex-valued function. For any $x \in [a, b]$ we can write

$$f(x) = f_1(x) + if_2(x)$$

where $f_1(x)$ and $f_2(x)$ are real numbers, which gives us two real-valued functions $f_1 : [a, b] \to \mathbb{R}$ and $f_2 : [a, b] \to \mathbb{R}$.

(i) The integral $\int_a^b f(x)dx$ is defined by

$$\int_a^b f(x)dx = \int_a^b f_1(x)dx + i\int_a^b f_2(x)dx$$

(provided the two real-valued integrals on the right hand side exist).

(ii) The *derivative* $f'(x)$ is defined by

$$f'(x) = f_1'(x) + if_2'(x)$$

(provided the two real-valued derivatives on the right hand side exist.) ■

10.4.2 Example. Consider $f : \mathbb{R} \to \mathbb{C}$ given by

$$f(x) = (x + 2i)^2 \quad \text{for } x \in \mathbb{R}.$$

Then

$$f(x) = (x^2 - 4) + 4xi$$

so that here $f_1(x) = x^2 - 4$ and $f_2(x) = 4x$. Thus, by the definitions above,

$$\begin{aligned}\int_1^2 (x + 2i)^2 dx &= \int_1^2 (x^2 - 4)dx + i\int_1^2 4xdx\\ &= -\frac{5}{3} + 6i\end{aligned}$$

and

$$f'(x) = f_1'(x) + if_2'(x) = 2x + 4i.$$ ■

Many of the properties of integrals and derivatives with which you are familiar carry over to these integrals and derivatives of complex-valued functions.* In the next proposition we collect some of the properties we need below.

10.4.3 Proposition. *Let $a, b \in \mathbb{R}$ with $a < b$, $c, d_1, d_2 \in \mathbb{C}$, and $f, g, F : [a, b] \to \mathbb{C}$ be complex-valued functions. Then, provided the relevant integrals and derivatives below exist,*

(i) $\int_a^b cf(x)dx = c\int_a^b f(x)dx$;

(ii) $\int_a^b (f(x) + g(x))dx = \int_a^b f(x)dx + \int_a^b g(x)dx$;

(iii) [Linearity]

$$\int_a^b (d_1 f(x) + d_2 g(x))\, dx = d_1\int_a^b f(x)dx + d_2\int_a^b g(x)dx;$$

---

* The subject of Complex Analysis, which is often taught near the end of most undergraduate courses, covers integration and differentiation of more general functions $f : \mathbb{C} \to \mathbb{C}$. Because we expect that most of our readers will not yet have taken such a course, we have chosen to introduce in this section the special case $f : \mathbb{R} \to \mathbb{C}$, basing our development on material familiar from elementary (real) calculus courses. Of course, any readers familiar with complex analysis will find the material in this section easy special cases of what they already know.

(iv) [Fundamental Theorem of Calculus]
if $F'(x) = f(x)$ *then* $\int_a^b f(x) = F(b) - F(a)$;
(v) [Integration by Parts]

$$\int_a^b f(x)g'(x)dx = f(b)g(b) - f(a)g(a) - \int_a^b f'(x)g(x)dx.$$

Proof. First we do (i) in some detail. Let $f(x) = f_1(x) + if_2(x)$ and $c = c_1 + c_2 i$, where $f_1(x)$, $f_2(x)$, c_1 and c_2 are real. Then

$$cf(x) = (c_1 f_1(x) - c_2 f_2(x)) + i(c_1 f_2(x) + c_2 f_1(x))$$

and so, by Definition 10.4.1,

$$\int_a^b cf(x)dx = \int_a^b (c_1 f_1(x) - c_2 f_2(x))dx + i\int_a^b (c_1 f_2(x) + c_2 f_1(x))dx.$$

The right hand side of (i) is easily seen to be the same. Part (ii) follows similarly, while (iii) follows easily from (i) and (ii).

The remaining results also follow easily from Definition 10.4.1 and the corresponding properties of real-valued functions. Write the details of at least some of the other parts (see Exercises 10.4 #3 below). ■

10.4.4 Lemma. *Let* $r \in \mathbb{C}$ *and consider* $g : \mathbb{R} \to \mathbb{C}$ *given by* $g(u) = \mathrm{e}^{-ur}$ *for* $u \in \mathbb{R}$. *Then* $g'(u) = -r\mathrm{e}^{-ur}$ *for all* $u \in \mathbb{R}$.

Proof. If $r = r_1 + r_2 i$ (r_1, r_2 real) then, from Definition 10.4.1,

$$g(u) = \mathrm{e}^{-ur_1}(\cos(-ur_2) + i\sin(-ur_2))$$

and the result follows easily from the definition of derivative in Definition 10.4.1. ■

We also need the following lemma which gives an upper bound for the absolute value of an integral. (Recall that the *absolute value* $|z|$ of the complex number $z = c + di$ is given by $|z| = \sqrt{c^2 + d^2}$.)

10.4.5 Lemma. *Let* $a, b, c \in \mathbb{R}$ *with* $a < b$ *and let* $f : [a, b] \to \mathbb{C}$ *be such that* $|f(x)| \le c$ *for all* $x \in [a, b]$. *Then*

$$\left|\int_a^b f(x)dx\right| \le 2c(b - a).$$

Proof. Let $f(x) = f_1(x) + if_2(x)$, where $f_1(x)$ and $f_2(x)$ are real. Then, if $x \in [a, b]$,

$$|f_1(x)| \le \sqrt{f_1(x)^2 + f_2(x)^2} = |f(x)| \le c$$

and similarly $|f_2(x)| \leq c$. Now, from (i) of Definition 10.4.1,

$$\begin{aligned}
\left|\int_a^b f(x)dx\right| &= \sqrt{\left(\int_a^b f_1(x)dx\right)^2 + \left(\int_a^b f_2(x)dx\right)^2} \\
&\leq \left|\int_a^b f_1(x)dx\right| + \left|\int_a^b f_2(x)dx\right| && (*) \\
&\leq c(b-a) + c(b-a) && (**) \\
&= 2c(b-a)
\end{aligned}$$

as required. (Note that $(*)$ follows because, as is easily seen by squaring, $\sqrt{r^2+s^2} \leq |r| + |s|$ for all $r, s \in \mathbb{R}$, while $(**)$ follows because $|f_1(x)| \leq c$ and $|f_2(x)| \leq c$.) ■

Exercises 10.4

1. Evaluate the complex-valued integral $\int_0^1 (x+i)^3 dx$.
2. If $f(x) = (x+i)^3$ for all $x \in \mathbb{R}$, use (ii) of Definition 10.4.1 to calculate $f'(x)$.
3. Complete the proof of Proposition 10.4.3, being careful to indicate which properties of integrals of real-valued functions you use.
4. Complete the proof of Lemma 10.4.4.
5. Let $f : \mathbb{R} \to \mathbb{C}$ be given by $f(n) = u\mathrm{e}^{-ui}$ for all $u \in \mathbb{R}$. Use integration by parts to calculate $\int_0^1 f(u)du$.

10.5 π is Transcendental – Part 2

Proposition 10.3.3 is the beginning of our proof that π is transcendental. Recall that we proved there that, if π is algebraic, then there exist integers $m \geq 1, q \geq 1, b \geq 1$ and nonzero complex numbers $\beta_1, \beta_2, \ldots, \beta_m$ such that

$$\mathrm{e}^{\beta_1} + \mathrm{e}^{\beta_2} + \ldots + \mathrm{e}^{\beta_m} + q = 0 \tag{1}$$

and the polynomial

$$h(X) = b(X - \beta_1)(X - \beta_2)\ldots(X - \beta_m) \tag{2}$$

has integer coefficients.

Expanding the right hand side of (2) gives

$$h(X) = b_0 + b_1 X + \ldots + b_{m-1}X^{m-1} + bX^m \tag{3}$$

where $b_0, \ldots, b_{m-1}, b$ are integers and the constant term

$$b_0 = (-1)^m b\beta_1\beta_2 \dots \beta_m$$

is nonzero since the b and β's are all nonzero.

Now that the above equations have been obtained, the rest of our proof that π is transcendental is going to look very much like the proof for e given in Section 10.2. We are going to derive a contradiction from (1) just as we derived a contradiction from the corresponding result ((1) of Section 10.2) for e. To this end we shall show how to construct certain numbers M, $M_{\beta_1}, \dots, M_{\beta_m}$ and $\varepsilon_{\beta_1}, \dots, \varepsilon_{\beta_m}$ by using integrals, as we did in Section 10.2 in the proof for e.

The integrals we shall use have the form

$$\int_0^1 f(ur)\mathrm{e}^{-ur} r du \qquad (*)$$

where r is a complex number. Interested readers might like to compare this with the integrals

$$\int_0^r f(x)\mathrm{e}^{-x} dx \quad (r \in \mathbb{R}) \qquad (**)$$

we used in Section 10.2. Although it is not used in our proof below, you may be interested to note that if we make the formal substitution $x = ur$ in $(**)$ we obtain Lemma 10.5.1(a).

Producing the M's and the ε's

We state and prove below four lemmas needed to define the M's and the ε's and to derive their properties. Each of these lemmas is very similar to one of the lemmas in Section 10.2.

The first lemma is the analogue of Lemma 10.2.1.

10.5.1 Lemma. *For each integer $k \geq 0$ there are polynomials $g_k(X)$, $h_k(X)$ with real coefficients such that, for all $r \in \mathbb{C}$,*

$$(a) \qquad \int_0^1 (ur)^k \mathrm{e}^{-ur} r du = k! - \mathrm{e}^{-r} g_k(r),$$

$$(b) \qquad \int_0^1 (ur - r)^k \mathrm{e}^{-ur} r du = h_k(r) - \mathrm{e}^{-r} k!.$$

Proof. This can be proved as for Lemma 10.2.1 by mathematical induction on k, using the relevant integration by parts rule (see (v) of Proposition 10.4.3). Lemma 10.4.4 gives the rule needed for differentiating (or antidifferentiating) e^{-ur} with respect to u. ∎

The next lemma is the analogue of Lemma 10.2.3.

10.5.2 Lemma. *Given a polynomial $f(X)$ with real coefficients, there is a unique number M and a unique polynomial $G(X)$ with real coefficients such that, for all $r \in \mathbb{C}$,*

$$\int_0^1 f(ur)\mathrm{e}^{-ur} r\,du = M - \mathrm{e}^{-r} G(r).$$

Indeed M and $G(X)$ are unique in the strong sense that, if $P_1(X)$ and $P_2(X)$ are polynomials with real coefficients such that

$$\int_0^1 f(ur)\mathrm{e}^{-ur} r\,du = P_1(r) - \mathrm{e}^{-r} P_2(r)$$

for all $r \in \mathbb{C}$, then $P_2(X) = G(X)$ and $P_1(X) = M$.

Proof. The proof is very similar to that of Lemma 10.2.3. For the existence part, use linearity as in (iii) of Proposition 10.4.3 instead of (i) of Theorem 10.1.2, and then use Lemma 10.5.1(a) instead of Lemma 10.2.1(a). The uniqueness part follows the proof in Lemma 10.2.3 exactly since in this case

$$M - P_1(r) = \mathrm{e}^{-r} \left(G(r) - P_2(r)\right)$$

holds for all $r \in \mathbb{C}$ and so, in particular, for all $r \in \mathbb{R}$. ■

As in the proof for e, our choice of the numbers M etc involves using a particular choice of polynomial in the integrand.

10.5.3 Definition. In Lemma 10.5.2 we choose $f(X)$ to be the polynomial

$$f(X) = \frac{X^{p-1}\left(h(X)\right)^p}{(p-1)!} \tag{4}$$

where the polynomial $h(X)$ is given by (2) and (3) and where p is a prime bigger than the integers q, b and $|b_0|$ occurring in (1), (2) and (3). We let $n = mp + p - 1$, which is the degree of $f(X)$.

We then choose

(i) M and $G(X)$ to be the number and polynomial given in Lemma 10.5.2,
(ii) $M_r = b^n G(r)$ for $r \in \{\beta_1, \ldots, \beta_m\}$, and
(iii) $\varepsilon_r = b^n \mathrm{e}^r \int_0^1 f(ur)\mathrm{e}^{-ur} r\,du$ for $r \in \{\beta_1, \ldots, \beta_m\}$. ■

Properties of $M, M_{\beta_1}, \ldots, M_{\beta_m}$ and $\varepsilon_{\beta_1}, \ldots, \varepsilon_{\beta_m}$

The next lemma is the analogue of Lemma 10.2.6.

10.5.4 Lemma. *(a) The number M is an integer which does not have the prime p as a factor when $p > |b_0|$, where b_0 is as in (3) above.*

(b) There is a polynomial $G_1(X)$ with integer coefficients and of degree at most n such that $G(r) = pG_1(r)$ for $r \in \{\beta_1, \ldots, \beta_m\}$.

Proof. The proof of each part (a) and (b) is very similar to the corresponding part of Lemma 10.2.6. Here we merely indicate the changes that are required and leave the details as Exercises 10.6 #3.

To prove (a), replace x by ur and dx by du, and use Lemmas 10.5.1(a) and 10.5.2 in place of Lemmas 10.2.1(a) and 10.2.3. Recall that $b_0 \neq 0$ in (3).

To prove (b), firstly expand $f(X)$ in powers of $(X - r)$, with $r \in \mathbb{C}$, using the obvious complex analogue of Proposition 10.1.6. Again replace x by ur and use Lemmas 10.5.1(b) and 10.5.2 in place of Lemmas 10.2.1(b) and 10.2.3. That each of $d_0(r), \ldots, d_{p-1}(r)$ is zero if r is in $\{\beta_1, \beta_2, \ldots, \beta_m\}$ follows just as in Lemma 10.2.5, using the complex analogues of Proposition 10.1.6 and Lemma 10.1.7. ∎

Note a significant difference between Lemma 10.2.6(b) and Lemma 10.5.4(b). In the latter we make no claim that $G_1(r)$ is an integer for $r \in \{\beta_1, \ldots, \beta_m\}$. Although $G_1(X)$ has integer coefficients, the β's are complex numbers and so we do not expect $G_1(r)$ to be an integer (unlike Section 10.2 where $r \in \{1, 2, \ldots, m\}$ so r is an integer, guaranteeing that $G_1(r) \in \mathbb{Z}$). In the proof for π we shall use symmetric polynomials to show that the sum

$$G_1(\beta_1) + \ldots + G_1(\beta_m)$$

is an integer (even though the individual $G_1(\beta_i)$'s may not be).

Note that a sequence $\{z_n\}$ of complex numbers converges to 0 as $n \to \infty$ if the sequences of real and imaginary parts of z_n converge to 0; this is equivalent to saying that the sequence $\{|z_n|\}$ of the absolute values converges to 0.

Our final lemma is the analogue of Lemma 10.2.7.

10.5.5 Lemma. *Let $f(X)$ and n be as in Definition 10.5.3. For $r \in \{\beta_1, \ldots, \beta_m\}$ let ε_r be as in Definition 10.5.3, so that*

$$\varepsilon_r = b^n \mathrm{e}^r \int_0^1 f(ur)\mathrm{e}^{-ur} r\,du.$$

Then

$$\lim_{p\to\infty} \varepsilon_r = 0.$$

Proof. Let $u \in [0, 1]$ and let $x = ur$ where r is the complex number $r = r_1 + ir_2$ (where $r_1, r_2 \in \mathbb{R}$). By definition,

$$\mathrm{e}^r = \mathrm{e}^{r_1}(\cos r_2 + i \sin r_2)$$

and so $|\mathrm{e}^r| = \mathrm{e}^{r_1}$. Similarly $|\mathrm{e}^{-ur}| = \mathrm{e}^{-ur_1} \leq \mathrm{e}^{|r_1|}$. Also, from (4) and (2),

$$
\begin{aligned}
|f(x)| &= \left|\frac{x^{p-1}}{(p-1)!}b^p(x-\beta_1)^p\dots(x-\beta_m)^p\right| \\
&\le \frac{|r|^{p-1}}{(p-1)!}b^p(|r|+|\beta_1|)^p\dots(|r|+|\beta_m|)^p .
\end{aligned}
$$

Hence, using Lemma 10.4.5,

$$
\begin{aligned}
&\left|b^n e^r \int_0^1 f(ur)e^{-ur} r du\right| \\
&\le b^{mp+p-1}e^{r_1}(2 \text{ length of interval}) \times (\text{upper bound for } |f(ur)e^{-ur}r|) \\
&\le b^{mp+p-1}e^{r_1}.2.1.\frac{|r|^{p-1}b^p}{(p-1)!}(|r|+|\beta_1|)^p\dots(|r|+|\beta_m|)^p e^{|r_1|}|r| \\
&\le \frac{2e^{r_1+|r_1|}}{b}\frac{C^p}{(p-1)!}
\end{aligned}
$$

where $C = b^{m+2}|r|(|r|+|\beta_1|)\dots(|r|+|\beta_m|)$.

Note that C is independent of p (and so are b and r_1). Hence by Lemma 10.1.4, $\lim_{p\to\infty} \varepsilon_r = 0$. ■

Given the above results, the long awaited proof of the transcendence of π is relatively straightforward.

10.5.6 Theorem. [Lindemann's Theorem] *π is a transcendental number.*

Proof. Suppose π is algebraic. We shall derive a contradiction.

We have already proved various consequences of this assumption and restated them at the start of this section. We let p be any prime which is larger than all of q, b and $|b_0|$ occurring in (1), (2) and (3), and define the polynomial $f(X)$ and the integer n as in (4) above.

It follows from Lemma 10.5.4 that there is an integer M not divisible by p and a polynomial $G_1(X)$ with integer coefficients and degree at most n such that, for $r \in \{\beta_1, \beta_2, \dots, \beta_n\}$,

$$
\int_0^1 f(ur)e^{-ur} r du = M - e^{-r} p G_1(r). \tag{5}
$$

We are now ready to introduce the numbers $M_{\beta_1}, \dots, M_{\beta_m}$ and $\varepsilon_{\beta_1}, \dots, \varepsilon_{\beta_m}$. Recall our Definition 10.5.3.

We will apply Lemma 10.5.4(b).

Let M and M_r, ε_r, for each r in $\{\beta_1, \beta_2, \ldots, \beta_m\}$, be as in Definition 10.5.3. From (iii) of Definition 10.5.3 and Lemma 10.5.4(b),

$$M_r = pb^n G_1(r) \tag{6}$$

and by (iii) of Definition 10.5.3,

$$\varepsilon_r = b^n \mathrm{e}^r \int_0^1 f(ur)\mathrm{e}^{-ur} r du. \tag{7}$$

Then, from (5), (6) and (7), we have

$$M_r + \varepsilon_r = pb^n G_1(r) + b^n \mathrm{e}^r (M - \mathrm{e}^{-r} pG_1(r)) = b^n \mathrm{e}^r M$$

and so

$$\mathrm{e}^r = \frac{M_r + \varepsilon_r}{Mb^n} \quad \text{for } r \in \{\beta_1, \beta_2, \ldots, \beta_m\}. \tag{8}$$

Unlike the case for e, here the numbers $M_{\beta_1}, \ldots, M_{\beta_m}$ may not be integers – indeed all that we can be sure of is that they are complex numbers since the β's and hence the r in (6) are complex. We only get the integer we need by adding them all up and then recognizing that there is a symmetric polynomial involved.

Now $G_1(X_1) + G_1(X_2) + \ldots + G_1(X_m)$ is clearly a symmetric polynomial in $X_1, X_2, \ldots, X_m$ and so it follows from the Fundamental Theorem on Symmetric Polynomials 8.3.17 that

$$G(X_1) + G(X_2) + \ldots + G(X_m) = H(\sigma_1, \sigma_2, \ldots, \sigma_m)$$

where H is a polynomial with integer coefficient and degree at most n. Substituting β_i for $X_i (i = 1, 2, \ldots, m)$ we see that

$$G_1(\beta_1) + G_1(\beta_2) + \ldots + G_1(\beta_m) = H(\mu_1, \mu_2, \ldots, \mu_m) \tag{9}$$

where μ_i is the i-th elementary symmetric function σ_i evaluated at the point $(\beta_1, \beta_2, \ldots, \beta_m)$. But, by (2) and (3) and Remark 8.3.15,

$$\mu_i = \pm b_{m-i}/b.$$

Since the degree of H is at most n, each term on the right hand side of (9) is equal to an integer divided by some power (no bigger than n) of b (since $b_0, b_1, \ldots, b_{m-1}$

and b are all integers). Thus

$$G_1(\beta_1) + G_1(\beta_2) + \ldots + G_1(\beta_m) = d/b^n$$

for some integer d, and so, from (6),

$$M_{\beta_1} + M_{\beta_2} + \ldots + M_{\beta_m} pb^n(d/b^n) = pd$$

is an integer divisible by p.

Now we know that this sum is an integer with p as a factor. So we can choose p large enough to get a contradiction.

This is done in essentially the same way as in the proof for e.

Substituting (8) into (1) gives

$$[qMb^n + M_{\beta_1} + \ldots + M_{\beta_m}] + [\varepsilon_{\beta_1} + \ldots + \varepsilon_{\beta_m}] = 0. \qquad (10)$$

But, since the integer M is not divisible by the prime p and since p is larger than the integers q and $|b|$, it follows that qMb^n is an integer not divisible by p. This means that the term in the first [ ] of (10) is an integer not divisible by p and hence is a nonzero integer. On the other hand, by Lemma 10.5.5, the absolute value of each term inside the second [ ] of (10) can be made arbitrarily small by choosing p large enough, and so the absolute value of this second [ ] term in (10) can be made less than $\frac{1}{2}$ (for example) by choosing p large enough. This is the contradiction we promised.

Hence π is transcendental. ■

At long last we are able to complete the proof that Problem III of the Introduction.

10.5.7 Corollary. *Squaring the circle is not possible with straightedge and compass.*

Proof. This follows immediately from Theorem 10.5.6 and the remarks in Section 6.2.3. ■

Exercises 10.5

1. Complete the details of the proof of Lemma 10.5.1, being careful to check which properties from Section 10.4 you use.
2. Repeat Exercise 1 for Lemma 10.5.2.
3. Complete the details of the proof of Lemma 10.5.4. State and prove the complex analogues of Propositions 10.1.6 and Lemma 10.1.7 which you need. Prove the analogue of Lemma 10.2.5 which you need.

10.6 Transcendental Number Theory

Transcendental Number Theory is a rich and active area of research. This section is but a small introduction. Hopefully it will whet your appetite to learn more.

There are very strong generalizations of Lindemann's Theorem 10.5.6 that π is transcendental. In this section we shall state, but not prove, some of these.

Lindemann proved in 1882 that e^{α} is transcendental for every non-zero algebraic number α. This implies that $i\pi$ is transcendental since $\mathrm{e}^{i\pi} = -1$. As i is algebraic, it follows that π is a transcendental number.

The German mathematician Karl Theodor Wilhelm Weierstrass (1815–1897) in Weierstrass (1885) proved the more general Theorem 10.6.1.

10.6.1 Theorem. [Lindemann-Weierstrass Theorem] *Given any distinct algebraic numbers $\alpha_1, \alpha_2, \ldots \alpha_n$, for $n \in \mathbb{N}$, the numbers $e^{\alpha_1}, e^{\alpha_2}, \ldots, e^{\alpha_n}$ are linearly independent over the field $\mathbb{A}$ of all algebraic numbers.*

Theorem 10.6.1 implies Lindemann's result that e^{α} is transcendental for every algebraic number $\alpha \neq 0$, from which one deduces Lindemann's Theorem 10.5.6.

The Canadian-American mathematician Ivan Morton Niven (1915–1999) shows in Niven (1967, §4 of Chapter 9) how to derive the following beautiful corollary.

10.6.2 Corollary. *The following numbers are transcendental:*

(i) e, π;

(ii) e^{α}, $\sin\alpha$, $\cos\alpha$, $\tan\alpha$, $\sinh\alpha$, $\cosh\alpha$, $\tanh\alpha$, *for any non-zero algebraic number* α;

(iii) $\log_e\alpha$, $\arcsin\alpha$ *and in general the inverse functions of those listed in (ii), for any non-zero algebraic number* α; *wherever multiple values are involved, every such value is transcendental.*

In 1900 the German mathematician David Hilbert (1862–1943) at the International Congress of Mathematicians in Paris posed 23 problems which guided much of the research in mathematics in the twentieth century. (For a discussion of these see Browder (1976a), Browder (1976b).)

Hilbert's 7th Problem. If α is an algebraic number such that $\alpha \neq 0, 1$ and β is an algebraic number which is not a real rational number (that is, $\beta \in \mathbb{A} \setminus \mathbb{Q}$), is α^{β} necessarily a transcendental number.

Hilbert regarded this problem as one of the hardest and certainly harder than proving (or disproving) the Riemann Hypothesis (which is still unproved today). However it was solved independently in 1934 by the Soviet mathematician Alexander Osipovich Gelfond (1906–1968) and by the German mathematician Theodor Schneider (1911–1988).

To state this result and others we introduce some terminology which makes stating and understanding the results easier.

10.6.3 Definition. Let $\mathcal{L}$ be the set of logarithms to base e of the set $\mathbb{A} \setminus \{0\}$ of non-zero algebraic numbers; that is

$$\mathcal{L} = \{\lambda \in \mathbb{C} : \mathrm{e}^{\lambda} \in \mathbb{A}\}. \qquad \blacksquare$$

Note that the members of $\mathbb{Q}$ are real numbers, while the members of $\mathbb{A}$ are complex numbers but are not necessarily real numbers.

$$\mathbb{Q} \subset \mathbb{R} \subset \mathbb{C}, \quad \mathbb{A} \not\subset \mathbb{R}, \quad \mathbb{Q} \subset \mathbb{A} \subset \mathbb{C}, \quad \mathbb{Q} \subset \mathbb{A} \cap \mathbb{R}.$$

Note also that $\mathcal{L}$ is a vector subspace of $\mathbb{C}$ over the field $\mathbb{Q}$ (but not over the field $\mathbb{A}$).

As mentioned above, Lindemann proved that e^{α} is transcendental for every non-zero algebraic number. In this new notation this becomes:

10.6.4 Theorem. *Each non-zero member of $\mathcal{L}$ is transcendental.*

10.6.5 Theorem. [Gelfond–Schneider Theorem] *If the complex numbers $\lambda_1, \lambda_2 \in \mathcal{L}$ are linearly independent over $\mathbb{Q}$, then they are linearly independent over $\mathbb{A}$.*

It is straightforward to show that the Gelfond–Schneider Theorem 10.6.5 is equivalent to the following theorem. [Hint. Use Exercises 10.6 #1.]

10.6.6 Theorem. *If the complex numbers $\lambda_1, \lambda_2 \in \mathcal{L}$, $\lambda_2 \neq 0$, then $\dfrac{\lambda_1}{\lambda_2}$ is either a transcendental number or a rational number.*

10.6.7 Remark. Let us see why the Gelfond–Schneider Theorem 10.6.5 or equivalently Theorem 10.6.6 gives a positive answer to Hilbert's 7th problem.

Suppose the answer to Hilbert's 7th problem is no; that is, there exist algebraic numbers α and β, $\alpha \neq 0, 1$, and $\beta \notin \mathbb{Q}$ such that α^{β} is an algebraic number. Put $\beta = \dfrac{\log\gamma}{\log\alpha}$, so that $\gamma = \alpha^{\beta}$. With $\lambda_1 = \log\gamma$ and $\lambda_2 = \log\alpha$, we see that Theorem 10.6.6 implies that $\beta = \dfrac{\lambda_1}{\lambda_2}$ is either a transcendental number or a rational number. This is a contradiction. So our supposition is false. Hence the Gelfond–Schneider Theorem 10.6.5 provides a positive answer to Hilbert's 7th problem. $\blacksquare$

10.6.8 Example. We use the positive answer to Hilbert's 7th problem to observe the following.

(i) $2^{\sqrt{2}}$ and $\sqrt[n]{2^{\sqrt{2}}}$, where $n \in \mathbb{N} \setminus \{1, 2\}$, are transcendental numbers.
$2^{\sqrt{2}}$ is known as the *Gelfond–Schneider constant* or *Hilbert number*. The Russian mathematician Rodion Osievich Kuzmin (1891–1949) first proved in 1930 that $2^{\sqrt{2}}$ is a transcendental number. It equals
2.6651441426902251886502972498731398 . . . .

(ii) Note that $\mathrm{e}^{\pi} = (\mathrm{e}^{i\pi})^{-i} = (-1)^{-i}$ and so e^{π} is a trancendental number. This number is known as *Gelfond's constant*.
It equals 23.140692632779269005729086367948547 . . . .

(iii) Noting that i is algebraic and $i \notin \mathbb{Q}$, we see that i^i is a transcendental number. It equals 0.20787957635076190854695561983497877 . . . . ■

10.6.9 Remark. At the time of writing, it is not known which of the following numbers are transcendental:

(i) π^{e};
(ii) $\pi + \mathrm{e}$;
(iii) $\pi - \mathrm{e}$;
(iv) $\frac{\pi}{\mathrm{e}}$;
(v) $\mathrm{e}.\pi$;
(vi) π^{π};
(vii) e^{e};
(viii) $\pi^{\sqrt{2}}$. ■

Using ideas in Exercises 2.1 #9 and Exercises 10.6 #5, we see that not all of the above numbers are algebraic.

We have seen the wonderful development of transcendental number theory with Lindemann's proof that π is transcendental in 1882, Hilbert's 7th problem in 1900 and Gelfond and Schneider's answer to Hilbert's 7th problem in 1934 and subsequent extensions in the first half of the twentieth century.
Giant steps forward since then were largely due to the English mathematician Alan Baker (1939–2018). In particular his generalizations of the Gelfond–Schneider Theorem in the 1960s, and his beautiful book on Transcendental Number Theory in 1975 and the revised edition in 1979. The back cover of his 1979 book says "*First published in 1975, this classic book gives a systematic account of transcendental number theory, that is those numbers which cannot be expressed as the roots of algebraic equations having rational coefficients. Their study has developed into a fertile and extensive theory enriching many branches of pure mathematics. Expositions are presented of theories relating to linear forms in the logarithms of algebraic numbers, of Schmidt's generalization of the Thue-Siegel-Roth theorem, of Shidlovsky's work on Siegel's E-functions and of Sprindzuk's solution to the Mahler conjecture. The volume was revised in 1979, however Professor Baker has taken this further opportunity to update the book including new advances in the theory and many new references.*"

Gelfond was not satisfied with the Gelfond–Schneider Theorem 10.6.5 and wanted it to be generalized from λ_1, λ_2 to $\lambda_1, \lambda_2, \ldots, \lambda_n$, for $n \in \mathbb{N}$. Indeed in 1960 Gelfond said

. . . *one may assume* . . . *that the most pressing problem in the theory of transcendental numbers is the investigation of the measures of transcendence of finite sets of logarithms of algebraic numbers.*

Alan Baker solved this problem and it has had numerous applications to transcendence theory, algebraic number theory and Diophantine equations. He received the prestigious Fields Medal in 1970 for this work and his applications of it to Diophantine equations. The Qualitative version of Baker's Theorem is Theorem 10.6.10.

10.6.10 Theorem. [Baker's Theorem – Qualitative version] *If the complex numbers $\lambda_1, \lambda_2, \ldots, \lambda_n \in \mathcal{L}$, where $n \in \mathbb{N}$, are linearly independent over $\mathbb{Q}$, then they are linearly independent over $\mathbb{A}$.*

This is a weak version of Baker's Theorem in that Baker proved not only that $\lambda_1, \lambda_2, \ldots, \lambda_n$ are linearly independent over $\mathbb{A}$, that is for $\beta_1, \beta_2, \ldots, \beta_n \in \mathbb{A} \setminus \{0\}$, $\beta_1\lambda_1 + \beta_2\lambda_2 + \cdots + \beta_n\lambda_n$ does not equal zero, but that this quantity is bounded away from zero in an effective way.

10.6.11 Definition. The set of complex numbers $\{\alpha_1, \alpha_2, \ldots, \alpha_n\}$, $n \in \mathbb{N}$ is said to be *algebraically independent* over a subfield $\mathbb{F}$ of $\mathbb{C}$ if $f(\alpha_1, \alpha_2, \ldots, \alpha_n) \neq 0$, for every nonzero polynomial $f(X_1, X_2, \ldots, X_n) \in \mathbb{F}[X_1, X_2, \ldots, X_n]$.

10.6.12 Example. (i) $\{\sqrt{\pi}\}$ is algebraically independent over $\mathbb{Q}$;
(ii) $\{3\pi - 1\}$ is algebraically independent over $\mathbb{Q}$;
(iii) $\{\sqrt{\pi}, 3\pi - 1\}$ is not algebraically independent over $\mathbb{Q}$ since the nonzero polynomial $f(x, y) = 3x^2 - y - 1$ equals 0 when $x = \sqrt{\pi}$ and $y = 3\pi - 1$.

10.6.13 Remark. The following is an equivalent statement of the Lindemann-Weierstrass Theorem 10.6.1: *Let $\alpha_1, \alpha_2, \ldots \alpha_n \in \mathbb{C}$, for $n \in \mathbb{N}$, be algebraic numbers which are linearly independent over $\mathbb{Q}$. Then the set $\{e^{\alpha_1}, e^{\alpha_2}, \ldots, e^{\alpha_n}\}$ is algebraically independent over $\mathbb{Q}$.*

10.6.14 Remark. In mathematics a *Conjecture* is a statement which some mathematicians think is probably true but nobody has proved that it is true.

In transcendental number theory the most famous conjecture is known as Schanuel's Conjecture. It was made by the American mathematician Stephen H. Schanuel (1933–2014) in the 1960s. (See Lang (1966).)

Schanuel's Conjecture. *Let $\alpha_1, \alpha_2, \ldots \alpha_n \in \mathbb{C}$, for $n \in \mathbb{N}$, be linearly independent over $\mathbb{Q}$. Then there exists a subset with n elements of $\{\alpha_1, \alpha_2, \ldots, \alpha_n, e^{\alpha_1}, e^{\alpha_2}, \ldots, e^{\alpha_n}\}$ which is algebraically independent over $\mathbb{Q}$.*

If Schanuel's Conjecture were true, then it would imply almost all the results we stated on transcendental numbers. It also would imply that all the numbers mentioned in Remark 10.6.9 are transcendental. For example it would imply that $\pi + e$ is transcendental. (See Exercises 10.6 #7.)

Exercises 10.6

1. Let $\mathbb{F}$ be a subfield of $\mathbb{C}$, $\lambda_1, \lambda_2 \in \mathbb{F}$, and $\lambda_2 \neq 0$. Prove that λ_1 and λ_2 are linearly independent over $\mathbb{F}$ if and only if $\dfrac{\lambda_1}{\lambda_2} \notin \mathbb{F}$.

2. Verify that $\dfrac{\log 2}{\log 3} \notin \mathbb{Q}$. Deduce from this and Theorem 10.6.6 that $\dfrac{\log 2}{\log 3}$ is a transcendental number.

3. Verify that $\dfrac{\log m}{\log n}$, where $m, n \in \mathbb{N}$ with p a prime number which divides m but not n, is a transcendental number.
[Hint. See Exercise #2 above.]

4. Let $\alpha_1 = \dfrac{m_1}{n_1}$ and $\alpha_2 = \dfrac{m_2}{n_2}$, where $m_1, n_1, m_2, n_2 \in \mathbb{N}$, and p a prime number which divides m_1 but p does not divide m_2 or n_1. Verify that $\dfrac{\log \alpha_1}{\log \alpha_2}$ is a transcendental number.
5. Let a and b be transcendental numbers. Prove that at least one of ab and $a + b$ is transcendental.
   [Hint. Consider the polynomial $x^2 - (a + b)x + ab = (x - a)(x - b)$ and observe that its coefficients are algebraic numbers. Then use Theorem 4.4.8 which says that the field $\mathbb{A}$ of algebraic numbers is algebraically closed.]
6. Find an example to show that Exercise 5 above would be false if we replaced "transcendental" everywhere by "irrational".

7.* Let $\alpha_1 = 1$ and $\alpha_2 = i\pi$.

   (i) Using the fact that π is transcendental, verify that α_1 and α_2 are linearly independent over $\mathbb{Q}$.
   (ii) Assuming Schanuel's Conjecture is true, show that there is a subset of $\{1, i\pi, e^1, e^{i\pi}\} = \{1, -1, i\pi, e\}$ with 2 elements which is algebraically independent over $\mathbb{Q}$.
   (iii) Show that (ii) implies that the set $\{e, i\pi\}$ is algebraically independent over $\mathbb{Q}$.
   (iv) Deduce from (iii) that $e + i\pi$ and $e - \pi$ are transcendental.
   Warning: Note that the conclusions in (iv) rely upon Schanuel's Conjecture which has not been proved true by mathematicians.

Additional Reading for Chapter 10

The idea of using integration to prove that e is transcendental was introduced in the original proof by the French mathematician Charles Hermite (1822–1901) in Hermite (1873). This proof for e was extended by the German mathematician (Carl Louis) Ferdinand von Lindemann (1852–1939) in Lindemann (1882) to give a proof that π is transcendental. Simplified versions of the original proofs have been given by many authors; most of these use the polynomials $f(X)$ given in Definitions 10.2.4 and 10.5.3. They generally use results about symmetric polynomials in the proofs for π.

Two variants of the proof stem from the German mathematicians David Hilbert (1862–1943) in Hilbert (1893) on the one hand, and from Adolf Hurwitz (1859–1919) in Hurwitz (1893) on the other. Although Hurwitz gives only the proof for e, his method was extended later to π by the Canadian-American mathematician Ivan Morton Niven 1915–1999) in Niven (1939). In Hilbert's approach, the integral of $f(x)\mathrm{e}^{-x}$ is obtained by first expanding the polynomial $f(x)$ in suitable powers, and then integrating term by term. This seems essentially easier than Hurwitz's approach in which the integral is expressed as the sum of the successive derivatives of $f(x)$, left in factorized form. In the proof for π, however, Niven's approach is more elementary than Hilbert's in that it avoids use of integration in the complex plane (and the Cauchy integral theorem). Our proof is an attempt to combine the advantages of both of these approaches.

References which follow Hilbert's approach are Baker (1975), Klein (1948) and Spivak (1967) – although the last gives only the proof for e. References which follow the Hurwitz-Niven approach include Hadlock (1978), Niven (1967) and Niven (1939).

In an attempt to make the proof even more elementary, a number of authors have removed integrals entirely from their proofs and replaced them by sums, as in Gordan (1893), Hobson (1953), Hardy (1952), Klein (1962) and Smith (1955). The American mathematician Michael David Spivak remarks that such elementary proofs are often harder to understand than the original proofs, inasmuch as the essential structure of the proof has been obscured just to remove a few integral signs. What do readers think?

An approach to symmetric polynomials which is similar to ours is contained in Ferrar (1958) and Hadlock (1978). A proof of the Fundamental Theorem on Symmetric Polynomials which uses a double induction, rather than the concept of order of a monomial, is given in Archbold (1970) and Clark (1971).

Useful references on Transcendental Number Theory are Baker (1975); Mahler (1976); Murty and Rath (2014); Natarajan and Thangadurai (2020) and Parshin (1997).

For background on Hilbert's seventh problem, see Garcia and Miller (2019).

To read a very recent paper on Topological Transcendental Fields, see Chalebgwa and Morris (2022). For an extension of Corollary 10.6.2, where "non-zero algebraic number" is replaced everywhere by "Liouville number", see Chalebgwa and Morris (n.d.).

Chapter 11
An Algebraic Postscript

Now we introduce some purely algebraic results which extend those in earlier chapters. These new results will enable you to construct fields that you have not seen before.

If α is a complex number which is algebraic over a subfield $\mathbb{F}$ *of* $\mathbb{C}$, we have been able to show that $\mathbb{F}(\alpha)$ is a field, but we have, as yet, introduced no effective way of calculating reciprocals of numbers in $\mathbb{F}(\alpha)$. In this chapter, we describe a way of constructing fields which gives, as a by-product, an algorithm for calculating such reciprocals.

The material in this chapter is optional as it is, to some extent a digression from our main theme of "impossibilities". We return to this theme in the short concluding Chapter 12.

11.1 The Ring $\mathbb{F}[X]_{p(X)}$

We describe a set $\mathbb{F}[X]_{p(X)}$ of polynomials and define addition and multiplication operations for these polynomials. It will turn out that $\mathbb{F}[X]_{p(X)}$ is a field with these operations.

11.1.1 Definitions. Let $\mathbb{F}$ be any field and let $p(X)$ be a polynomial in $\mathbb{F}[X]$ which is irreducible over $\mathbb{F}$ and which has degree n. Define the set $\mathbb{F}[X]_{p(X)}$ by putting

$$\mathbb{F}[X]_{p(X)} = \{a_0 + a_1 X + \ldots + a_{n-1} X^{n-1} : a_0, a_1, \ldots, a_{n-1} \in \mathbb{F}\}.$$

We also define addition and multiplication in $\mathbb{F}[X]_{p(X)}$: for each $f(X)$ and $g(X)$ in $\mathbb{F}[X]_{p(X)}$, define

(i) $f(X) +_{p(X)} g(X) = f(X) + g(X)$,

(ii) $f(X) \times_{p(X)} g(X) =$ remainder on dividing $f(X)g(X)$ by $p(X)$. ■

S. A. Morris et al., *Abstract Algebra and Famous Impossibilities*, Undergraduate Texts in Mathematics, https://doi.org/10.1007/978-3-031-05698-7_11

Notice that, as a set, $\mathbb{F}[X]_{p(X)}$ depends only on the degree n of the polynomial $p(X)$. The set $\mathbb{F}[X]_{p(X)}$ contains all polynomials in $\mathbb{F}[X]$ of degree less than n (together with the zero polynomial). Addition $+_{p(X)}$ in $\mathbb{F}[X]_{p(X)}$ is just ordinary addition of polynomials while multiplication $\times_{p(X)}$ depends on the polynomial $p(X)$. Indeed, multiplication is done *modulo* $p(X)$ in the same way that multiplication $\times_n$ in $\mathbb{Z}_n$ is done modulo n (as described in Section 1.1). The usual product $f(X)g(X)$ may have too high a degree for it to belong to $\mathbb{F}[X]_{p(X)}$; to bring down the degree we subtract multiples of $p(X)$. We can define $f(X)g(X)(\bmod\ p(X))$ to be $f(X)g(X)$ reduced to a degree less than the degree of the polynomial $p(X)$ by subtracting multiples of $p(X)$.

11.1.2 Example. In the set of polynomials $\mathbb{Q}[X]_{X^3-2}$ we have

$$(X^2+3)+_{(X^3-2)}(2X+1)=X^2+2X+4$$

and

$$(X^2+3)\times_{(X^3-2)}(2X+1)=X^2+6X+7$$

since $(X^2+3)(2X+1)=2X^3+X^2+6X+3$ which leaves a remainder of X^2+6X+7 when divided by X^3-2. ■

11.1.3 Proposition. $(\mathbb{F}[X]_{p(X)}, +_{p(X)}, \times_{p(X)})$ *is a ring.*

Proof. Addition is just ordinary addition while multiplication is closely related to ordinary multiplication. The proof of the various ring axioms follows closely the arguments which one would use to prove that $(\mathbb{Z}_n, +_n, \times_n)$ is a ring. The details are left for you to consider. ■

In fact $\mathbb{F}[X]_{p(X)}$ is a ring, even if $p(X)$ is reducible over $\mathbb{F}$. The irreducibility of $p(X)$ over $\mathbb{F}$ is only needed (as we shall see in the next section) to prove that $\mathbb{F}[X]_{p(X)}$ is a field. This is also analogous to $\mathbb{Z}_n$, which is a ring, but only a field if n is a prime number.

Notice that all the constant polynomials are in $\mathbb{F}[X]_{p(X)}$ so that $\mathbb{F}\subseteq\mathbb{F}[X]_{p(X)}$. Indeed $\mathbb{F}[X]_{p(X)}$ is obviously a vector space over $\mathbb{F}$ with a basis $\{1, X, \ldots, X^{n-1}\}$ and dimension n.

Exercises 11.1

1. In $\mathbb{Q}[X]_{X^3-2}$, calculate
   (a) $(X^2-3X+2)+_{(X^3-2)}(2X^2+X+4)$,
   (b) $(X^2-3X+2)\times_{(X^3-2)}(2X^2+X+4)$.
2. (a) Explain why X^2+1 is irreducible over $\mathbb{R}$.
   (b) In $\mathbb{R}[X]_{X^2+1}$, calculate

(i) $(3+4X)+_{(X^2+1)}(2-X)$, (ii) $(3+4X)\times_{(X^2+1)}(2-X)$,
(iii) $(a+bX)+_{(X^2+1)}(c+dX)$, (iv) $(a+bX)\times_{(X^2+1)}(c+dX)$.

(c) Do the formulae in (b) (iii), (iv) look familiar? (If not, write down formulae for addition and multiplication of two complex numbers.)

3. (a) Show that in the ring $\mathbb{Q}[X]_{X^3-1}$, the element $X-1$ is a zero divisor. [Hint. Write X^3-1 as a product of two polynomials of smaller degree, and then multiply those polynomials modulo X^3-1.]
   (b) Is X^3-1 irreducible over $\mathbb{Q}$? Is $\mathbb{Q}[X]_{X^3-1}$ a field?

11.2 Division and Reciprocals in $\mathbb{F}[X]_{p(X)}$

Now that the set of polynomials $\mathbb{F}[X]_{p(X)}$ has been given the structure of a ring, it makes sense to ask whether this ring contains the reciprocal of each of its nonzero elements. A clever little algorithm can be used to show the existence of reciprocals, thereby giving an affirmative answer to our question.

The algorithm depends on the validity of the following lemma, which may seem paradoxical at first sight. It says that *you can reduce the degree of a polynomial by multiplying it by another polynomial*! You must remember, however, that it is not the usual multiplication of polynomials which is involved, but rather multiplication "modulo $p(X)$".

11.2.1 Lemma. [Degree Reducing Lemma] *Assume $a(X)$ is a nonconstant polynomial in $\mathbb{F}[X]_{p(X)}$. If $p(X)$ is irreducible over $\mathbb{F}$ then there exists a polynomial $q(X)$ in $\mathbb{F}[X]_{p(X)}$ such that the polynomial*

$$a(X)\times_{p(X)} q(X)$$

is nonzero and has degree less than that of $a(X)$.

Proof. By the Division Theorem 1.3.2, we can divide $p(X)$ by $a(X)$ to get a quotient $q(X)$ and remainder $r(X)$, both in $\mathbb{F}[X]$, such that

$$p(X)=a(X)q(X)+r(X) \qquad (*)$$

and either (i) $r(X)=0$
or (ii) $\deg r(X)<\deg a(X)$.

But if (i) holds then, by (*),

$$p(X)=a(X)q(X). \qquad (**)$$

Since, however, $a(X)$ is a nonconstant polynomial in $\mathbb{F}[X]_{p(X)}$,

$$0 < \deg a(X) < \deg p(X)$$

and so (**) contradicts the irreducibility of $p(X)$. Thus (i) is false and hence

$$r(X) \neq 0 \tag{1}$$

and (ii) holds. Thus

$$\deg r(X) < \deg a(X). \tag{2}$$

From (*) it now follows that

$$\begin{aligned} \deg p(X) &= \deg\, (a(X)q(X)) \\ &= \deg a(X) + \deg q(X). \end{aligned}$$

This shows that $\deg q(X) < \deg p(X)$ and hence

$$q(X) \in \mathbb{F}[X]_{p(X)}. \tag{3}$$

Finally note that, by (*), the remainder on dividing $a(X)q(X)$ by $p(X)$ is $-r(X)$. Hence

$$a(X) \times_{p(X)} q(X) = -r(X). \tag{4}$$

The conclusions of the lemma now follow from (1), (2), (3), and (4). ■

Note that there is an algorithm implicit in the above proof: *To get the "degree reducing factor" $q(X)$ you divide the irreducible polynomial $p(X)$ by the polynomial $a(X)$ whose degree you wish to reduce. The quotient $q(X)$ and the remainder $r(X)$ then satisfy*

$$a(X) \times_{p(X)} q(X) = -r(X)$$

where $r(X)$ is nonzero and has lower degree than $a(X)$.

We are now able to give an algorithm for calculating reciprocals in $\mathbb{F}[X]_{p(X)}$. For brevity, we use the symbol $\times$ to stand for the multiplication $\times_{p(X)}$ in $\mathbb{F}[X]_{p(X)}$.

11.2.2 Algorithm. [Calculation of Reciprocals] *Assume that $a(X)$ is a nonzero polynomial in the ring $\mathbb{F}[X]_{p(X)}$ where $p(X)$ is irreducible over $\mathbb{F}$. Construct, using the Degree Reducing Lemma 11.2.1, polynomials $q_1(X), q_2(X), q_3(X), \ldots$* in $\mathbb{F}[X]_{p(X)}$ *such that*

$$\begin{aligned} &a(X) \times q_1(X), \\ &a(X) \times q_1(X) \times q_2(X), \\ &a(X) \times q_1(X) \times q_2(X) \times q_3(X), \end{aligned}$$

and so on, have successively lower degrees. Hence eventually we get

$$a(X) \times q_1(X) \times q_2(X) \times q_3(X) \times \ldots \times q_m(X) = c,$$

where $c \neq 0$ is a constant in the field $\mathbb{F}$. If we put

$$d(X) = \frac{1}{c} q_1(X) \times q_2(X) \times q_3(X) \times \ldots \times q_m(X),$$

then $d(X)$ is the reciprocal of $a(X)$ in $\mathbb{F}[X]_{p(X)}$.

Proof. It is easy to see that

$$a(X) \times d(X) = 1.$$

■

The polynomial $d(X)$ so constructed, being the reciprocal of $a(X)$ in the ring $\mathbb{F}[X]_{p(X)}$, is denoted by

$$1/a(X)$$

(although this notation does not indicate the dependence of the answer on the polynomial $p(X)$).

The following simple description of the above algorithm for obtaining $1/a(X)$ may be helpful when you apply it to specific examples.

(A) *Find successive reducing factors for $a(X)$ until you get a constant.*

(B) *Multiply all the reducing factors together (modulo $p(X)$) and then divide by the constant. This gives you $1/a(X)$.*

11.2.3 Example. Find the reciprocal of $X^2 + 1$ in $\mathbb{Q}[X]_{X^3-2}$.

(A) *We find successive reducing factors for $X^2 + 1$.*

1. To find a reducing factor for $X^2 + 1$, divide this polynomial into $X^3 - 2$ to get

$$X^3 - 2 = (X^2 + 1)X + (-X - 2).$$

   Hence

$$(X^2 + 1) \times_{(X^3-2)} X = X + 2.$$

   Thus $X + 2$ is the result of applying the *reducing factor* X.

2. To find a reducing factor for $X + 2$, divide $X^3 - 2$ by $X + 2$ to get

$$X^3 - 2 = (X + 2)(X^2 - 2X + 4) - 10.$$

   Hence

$$(X + 2) \times_{(X^3-2)} (X^2 - 2X + 4) = 10.$$

Thus the constant 10 is the result of applying a further *reducing factor* $X^2 - 2X + 4$.

(B) *We multiply the reducing factors and divide by the constant.* This gives

$$\frac{1}{X^2+1} = \frac{1}{10}X \times_{(X^3-2)} (X^2 - 2X + 4)$$
$$= -\frac{1}{5}X^2 + \frac{2}{5}X + \frac{1}{5}.$$

■

It now follows easily that $\mathbb{F}[X]_{p(x)}$ is a field (provided $p(X)$ is irreducible over $\mathbb{F}$).

11.2.4 Theorem. *If $p(X)$ is irreducible over $\mathbb{F}$ then $\mathbb{F}[X]_{p(X)}$, with the operations in Definitions 11.1.1, is a field.*

Proof. We know from Section 11.1 that it is a ring. It is clearly commutative and contains 1. By Algorithm 11.2.2, it is closed under division by nonzero elements. Hence it is a field. ■

This gives us an important method for constructing new fields – find a polynomial $p(X)$ which is irreducible over a field $\mathbb{F}$ and construct $\mathbb{F}[X]_{p(X)}$. For example, you may have thought that there could be no field with four elements since the obvious contender $\mathbb{Z}_4$ is not a field. We can, however, use the $\mathbb{F}[X]_{p(X)}$ construction to exhibit a field with four elements.

11.2.5 Example. [A Field with Four Elements] We shall apply Theorem 11.2.4, with $\mathbb{F} = \mathbb{Z}_2$ and $p(X) = X^2 + X + 1$.

Because $p(0) \neq 0$ and $p(1) \neq 0$, it follows from the Small Degree Irreducibility Theorem 4.2.3 that $p(X)$ is irreducible over $\mathbb{Z}_2$. Now

$$\mathbb{E} = \mathbb{Z}_2[X]_{p(X)} = \{a_0 + a_1 X : a_0, a_1 \in \mathbb{Z}_2\}$$

has just four elements $0, 1, X, 1 + X$ and, by the above theorem, it is a field.

It is easy to draw up tables for the two operations in this field. The tables are shown below.

$+_{p(X)}$	0	1	X	$1+X$
0	0	1	X	$1+X$
1	1	0	$1+X$	X
X	X	$1+X$	0	1
$1+X$	$1+X$	X	1	0

$\times_{p(X)}$	0	1	X	$1+X$
0	0	0	0	0
1	0	1	X	$1+X$
X	0	X	$1+X$	1
$1+X$	0	$1+X$	1	X

Note also that (as you might expect from Theorem 3.2.4)

$$[\mathbb{E} : \mathbb{Z}_2] = \deg p(X),$$

each of these numbers being equal to 2. ■

Exercises 11.2

1. In the ring $\mathbb{Q}[X]_{X^3-2}$ find successive reducing factors for the polynomial $X^2 + X + 1$ and hence find the reciprocal of this element.
2. (a) In the ring $\mathbb{R}[X]_{X^2+1}$ find
   (i) $1/(3+4X)$,
   (ii) $1/(a+bX)$, if a and b are not both zero.
   (b) Does your answer to (a)(ii) agree with the formula for $1/(a+bi)$ in $\mathbb{C}$? (See part (c) of Exercises 11.1 #2.)
3. Which of the following are fields? (Justify your answers.)
   (a) $\mathbb{Q}[X]_{X^3-2X+1}$,
   (b) $\mathbb{Q}[X]_{X^3+2X+1}$,
   (c) $\mathbb{Q}(\sqrt[3]{2})[X]_{X^3-2}$.
4. In Example 11.2.3, check the very last step of the calculation and then verify directly, using multiplication modulo X^3-2, that the answer given for the reciprocal of $1+X^2$ is indeed correct.
5. Find a polynomial of degree 2 in $\mathbb{Z}_3[X]$ which is irreducible over $\mathbb{Z}_3$. Follow the method of Example 11.2.5 to construct a field with nine elements.
6. Show how to construct a field with eight elements.
   [Hint. You will need a polynomial of degree three in $\mathbb{Z}_2[X]$ which is irreducible over $\mathbb{Z}_2$.]

11.3 Reciprocals in $\mathbb{F}(\alpha)$

We return to the situation in the earlier chapters where α is a complex number which is algebraic over a subfield $\mathbb{F}$ of $\mathbb{C}$, with

$$\operatorname{irr}(\alpha, \mathbb{F}) = p(X) \quad \text{and} \quad \deg(\alpha, \mathbb{F}) = n.$$

Although we were able to prove in Chapter 3 that $\mathbb{F}(\alpha)$ is a field, we had no efficient way of expressing $1/\beta$ in the form

$$c_0 + c_1\alpha + \ldots + c_{n-1}\alpha^{n-1} \quad (c_i \in \mathbb{F})$$

for an arbitrary nonzero β in $\mathbb{F}(\alpha)$. We can now do this very easily, however, by using the algorithm from Section 11.2.

If you look at the definition of $\mathbb{F}(\alpha)$ (Definition 3.2.1) and at the definition of $\mathbb{F}[X]_{p(X)}$ in Section 11.1, you will notice a certain similarity. We expose this by defining the *evaluation mapping*

$$\phi_\alpha : \mathbb{F}[X]_{p(X)} \longrightarrow \mathbb{F}(\alpha)$$

by putting

$$\phi_\alpha(c_0 + c_1X + \ldots c_{n-1}X^{n-1}) = c_0 + c_1\alpha + \ldots + c_{n-1}\alpha^{n-1}.$$

Note that ϕ_α simply replaces the indeterminate X by the number α.

Recall that a field isomorphism is a one-to-one and onto mapping between two fields which preserves the addition and multiplication operations. Such an isomorphism also preserves subtraction and division. Two isomorphic fields are essentially the same, except for the names of their elements.

11.3.1 Proposition. *The map ϕ_α is a field isomorphism.*

Proof. That ϕ_α is a one-to-one and onto map follows routinely from the fact that $\{1, \alpha, \ldots, \alpha^{n-1}\}$ is a basis for $\mathbb{F}(\alpha)$ over $\mathbb{F}$ (see Exercises 11.3 #8). It is also easy to show that ϕ_α preserves addition in the sense that

$$\phi_\alpha\left(f(X) +_{p(X)} g(X)\right) = \phi_\alpha\left(f(X)\right) + \phi_\alpha\left(g(X)\right)$$

for all polynomials $f(X)$ and $g(X)$ in $\mathbb{F}[X]$.

What about multiplication? This is a bit more complicated. By the Division Theorem 1.3.2, there exist polynomials $q(X)$ and $r(X)$ in $\mathbb{F}[X]$ with

$$f(X)g(X) = p(X)q(X) + r(X) \tag{1}$$

where $r(X) = 0$ or $\deg r(X) < \deg p(X)$. If we replace X by α in (1) we get

$$\begin{aligned} f(\alpha)g(\alpha) &= p(\alpha)q(\alpha) + r(\alpha) \\ &= 0 \cdot q(\alpha) + r(\alpha) \\ &= r(\alpha). \end{aligned} \tag{2}$$

But by the definition of multiplication in the field $\mathbb{F}[X]_{p(X)}$,

$$r(X) = f(X) \times_{p(X)} g(X).$$

Hence in the equation (2),

$$\text{the left hand side} = \phi_\alpha(f(X)) \cdot \phi_\alpha(g(X))$$

while

$$\text{the right hand side} = \phi_\alpha\left(f(X) \times_{p(X)} g(X)\right).$$

These two sides are equal, which implies that the map ϕ_α preserves multiplication. ■

We can now put the algorithm for reciprocals in $\mathbb{F}[X]_{p(X)}$ to work in $\mathbb{F}(\alpha)$ since these two fields are isomorphic.

11.3.2 Example. Find the reciprocal of the number $1 + (\sqrt[3]{2})^2$ in the field $\mathbb{Q}(\sqrt[3]{2})$.

Note firstly that, because

$$\text{irr}(\sqrt[3]{2}, \mathbb{Q}) = X^3 - 2,$$

it follows that $\phi_{\sqrt[3]{2}}$ is an isomorphism from $\mathbb{Q}[X]_{X^3-2}$ to $\mathbb{Q}(\sqrt[3]{2})$. Under this isomorphism,

$$1 + (\sqrt[3]{2})^2 \in \mathbb{Q}(\sqrt[3]{2}) \text{ corresponds to } 1 + X^2 \in \mathbb{Q}[X]_{X^3-2}$$

or, more formally,

$$1 + (\sqrt[3]{2})^2 = \phi_{\sqrt[3]{2}}(1 + X^2).$$

Hence

$$\begin{aligned}
\frac{1}{1 + (\sqrt[3]{2})^2} &= \frac{1}{\phi_{\sqrt[3]{2}}(1 + X^2)} \\
&= \phi_{\sqrt[3]{2}}\left(\frac{1}{1 + X^2}\right) && \text{as } \phi_{\sqrt[3]{2}} \text{ is an isomorphism} \\
&= \phi_{\sqrt[3]{2}}\left(-\frac{1}{5}X^2 + \frac{2}{5}X + \frac{1}{5}\right) && \text{by Example 11.2.3} \\
&= -\frac{1}{5}(\sqrt[3]{2})^2 + \frac{2}{5}(\sqrt[3]{2}) + \frac{1}{5}.
\end{aligned}$$

■

This concludes our algebraic postscript in which we have shown how to construct new fields (including finite ones) and have given an efficient way of calculating reciprocals in fields of the form $\mathbb{F}[X]_{p(X)}$ and $\mathbb{F}(\alpha)$.

Exercises 11.3

1. For which choice of the polynomial $p(X)$ is the field $\mathbb{Q}(\sqrt[3]{2})$ isomorphic to the field $\mathbb{Q}[X]_{p(X)}$?
2. Check (using multiplication in $\mathbb{Q}(\sqrt[3]{2})$) the answer obtained in Example 11.3.2.
3. (a) Write out explicitly each of the sets $\mathbb{Q}(\sqrt[3]{2})$ and $\mathbb{Q}[X]_{X^3-2}$.
   (b) Write down (i) a typical element of the latter set,
   (ii) its image under the map $\phi_{\sqrt[3]{2}}$,
   and show both in the appropriate sets in Figure 11.1.

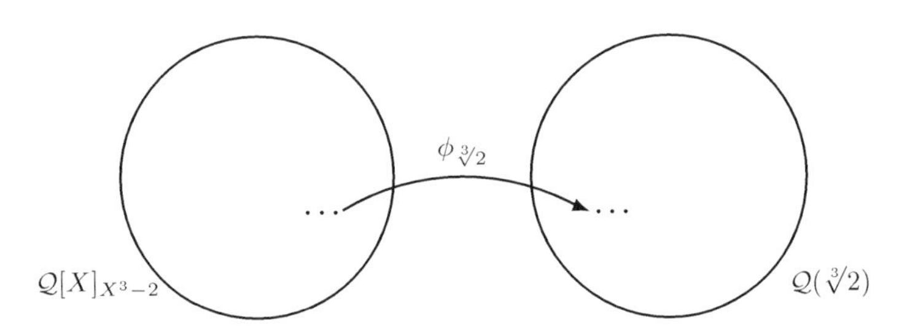

Fig. 11.1 The map $\phi_{\sqrt[3]{2}}$

   (c) Which theorem from Chapter 3 guarantees that each element of $\mathbb{Q}(\sqrt[3]{2})$ is *uniquely expressible* in the form

$$c_0 + c_1\sqrt[3]{2} + c_2(\sqrt[3]{2})^2$$

   with each c_i in $\mathbb{Q}$?
   (d) Apply the map $\phi_{\sqrt[3]{2}}{}^{-1}$ (the inverse of the map $\phi_{\sqrt[3]{2}}$) to the above element.
4. Let $f(X) = X^2 + 1$ and $g(X) = 2X + 3$, which are elements of $\mathbb{Q}[X]$.
   (a) State why $f(X)$ and $g(X)$ belong to $\mathbb{Q}[X]_{X^3-2}$.
   (b) Calculate the following sum and product in the ring $\mathbb{Q}[X]_{X^3-2}$.
   (i) $f(X) +_{(X^3-2} g(X)$,
   (ii) $f(X) \times_{(X^3-2))} g(X)$.
   (c) Use the fact that $\phi_{\sqrt[3]{2}}$ is a homomorphism from $\mathbb{Q}[X]_{X^3-2}$ to $\mathbb{Q}(\sqrt[3]{2})$ to calculate its value at the product in (b)(ii). Check your answer by applying the map directly to your answer for (b)(ii).
5. Use your answer to Exercises 11.2 #1 to find the reciprocal in $\mathbb{Q}(\sqrt[3]{2})$ of the number $(\sqrt[3]{2})^2 + \sqrt[3]{2} + 1$. Indicate where you use the fact that $\phi_{\sqrt[3]{2}}$ is an isomorphism.

6. Repeat Exercise 5, but this time follow the method used in the proof of Theorem 3.2.6 and in Example 3.2.7. Which method seems the more efficient, this one or that of Exercise 5?
7. Use your answer to Exercise 5 to express

$$\left(1+6\sqrt[3]{2}-2(\sqrt[3]{2})^2\right) / \left(1+\sqrt[3]{2}+(\sqrt[3]{2})^2\right)$$

in the form $c_0 + c_1\sqrt[3]{2} + c_2(\sqrt[3]{2})^2$ for $c_0, c_1, c_2 \in \mathbb{Q}$.
8. Prove that the map ϕ_α defined in the text is one-to-one and onto.
[Hint. Use the fact that $\{1, \alpha, \ldots, \alpha^{n-1}\}$ is a basis for $\mathbb{F}(\alpha)$ over $\mathbb{F}$.]

Additional Reading for Chapter 11

The field $\mathbb{F}[X]_{p(X)}$ is usually constructed as a factor ring of the ring $\mathbb{F}[X]$ of polynomials. See, for example, (Fraleigh, 1982, Section 35) and (Gilbert, 1976, Chapter 10).

Reciprocals of elements in $\mathbb{F}(\alpha)$ are often calculated using the algorithm for finding the greatest common divisor of two polynomials, as indicated, for example, (Clark, 1971, Section 110) and (Hadlock, 1978, Section 2.2).

Finite fields have been completely characterized: see, for example, (Fraleigh, 1982, Section 45) and (Gilbert, 1976, Chapter 11).

Chapter 12
Other Impossibilities: Regular Polygons and Integration in Finite Terms

This book has introduced some basic ideas from the part of abstract algebra which deals with fields. Use of these ideas made it relatively easy for us to prove the impossibilities mentioned in the Introduction.

It may now come to you as a pleasant surprise that your newly acquired ideas about fields can also be used to prove the impossibility of solving other famous problems. Two such problems are discussed below. While the first problem also comes from geometry, the other comes from elementary calculus.

12.1 Construction of Regular Polygons

A polygon is said to be *regular* if all its sides have the same length and all its angles are equal. Constructing a regular polygon with n sides amounts to dividing an angle of $360°$ into n equal parts.

Using only straightedge and compass the ancient Greeks were able to construct regular polygons with 3 sides and with 5 sides, but were not able to construct one with 7 sides. It was, of course, trivial for them to construct a polygon with 2^m sides, for any integer $m \geq 2$, as this merely involves repeated bisection. They also proved that if it is possible to construct regular polygons with r sides and s sides, then it is also possible to construct one with rs sides, provided that the integers r and s have greatest common divisor 1. Thus, taking $r = 3$ and $s = 5$, they knew it is possible to construct one with 15 sides.

No further progress was made on the problem of constructing regular polygons for over 2000 years until in 1796 the nineteen year old *Carl Friedrich Gauss* amazed the mathematical world by constructing a regular polygon with 17 sides. He did this by showing that the equation $x^{17} - 1 = 0$ can be reduced to a finite set of quadratic equations. So the regular polygon with 17 sides can be constructed with straightedge

S. A. Morris et al., *Abstract Algebra and Famous Impossibilities*, Undergraduate Texts in Mathematics, https://doi.org/10.1007/978-3-031-05698-7_12

and compass. Gauss showed, furthermore, that a regular polygon with n sides can be constructed whenever n is a prime number of the form

$$n = 2^{2^k} + 1$$

for some integer $k \geq 0$. Prime numbers of this form are called *Fermat primes*. (Notice that $17 = 2^{2^2} + 1$ is a Fermat prime.)

From the results mentioned above, it follows that a regular polygon with n sides can be constructed if

$$n = 2^{i+2} \quad \text{or} \quad n = 2^i p_1 p_2 \dots p_j$$

where $i \geq 0$ is an integer and $p_1, p_2, \dots, p_j$ are distinct Fermat primes. What if n does not have this form? The answer was provided by *Pierre Wantzel* in 1837, who proved that, for n not of this form, the construction of a regular polygon with n sides is impossible. The proof of this impossibility requires little more than the ideas about fields already given in this book. Books which cover this topic in detail are listed in the "Additional Reading" at the end of this chapter.

12.2 Integration in Closed Form

A topic which has tended to dominate the undergraduate curriculum in elementary calculus is that of finding indefinite integrals in "*closed form*"; that is, in terms of functions which are usually defined in introductory calculus courses.

This topic is usually presented as an art rather than as a science: students can evaluate an integral if they are clever enough (or lucky enough!) to guess a suitable trick. There are, however, some integrals which look quite innocent and yet refuse to "come out". One such example is

$$\int e^{x^2} dx.$$

The impossibility of finding this integral in closed form (also called in "*finite terms*") was proved by the French mathematician Joseph Liouville (1809–1882), who, in the years 1833–1841, laid the foundations for the study of integration as a science, rather than an art.

At first sight one might guess that, of all the topics in the undergraduate curriculum, none could be further removed from abstract algebra than that of finding integrals in closed form – not so, however! Spivak (1967) mentions that polished algebraic treatments of Liouville's theorems had recently been obtained.

You might well ask how is abstract algebra, in particular the field theory you have learned in this book, relates to calculus and more specifically to indefinite integrals. It took 140 years for the work of Liouville to be presented satisfactorily in algebraic terms. To see how this was done we introduce the topic of *differentiable fields*. Our aim is not to teach you about differentiable fields, but rather to show you how a problem on indefinite integration can be translated into a problem in field theory.

First we observe that if $\mathbb{F}$ is a field, then the set of polynomials $\mathbb{F}[X]$ forms a ring but not a field because it lacks division. If $p(X), q(X)$ are in $\mathbb{F}[X]$, then $f(X) = \dfrac{p(X)}{q(X)}$ is rarely a polynomial. So we extend $\mathbb{F}[X]$ to a field using rational functions.

12.2.1 Definition. Let $\mathbb{F}$ be a field and $p(X)$ and $q(X)$ polynomials in $\mathbb{F}[X]$. Then $f(X)$ is said to be a *rational function* if $f(X) = \dfrac{p(X)}{q(X)}$, where $f(X)$ is defined for all X such that $q(X) \neq 0$. ■

Now it is easy to define what we mean by differentiation and differentiable fields. The term *differentiable field* was introduced by the American mathematician Joseph Fels Ritt (1893–1951).

12.2.2 Definitions. Let R be a ring. Then a function $' : R \to R$ such that $f \mapsto f'$ is said to be a *derivation* if $(fg)' = f'g + fg'$, and $(f+g)' = f' + g'$, for all $f, g \in R$. The pair $(R, ')$ is said to be a *differentiable ring*. The element $f \in R$ is said to be a *constant* if $f' = 0$. The set of constants of R is denoted by $\text{const}(R, ')$. If R is a field $\mathbb{F}$, then the pair $(\mathbb{F}, ')$ is said to be a *differentiable field*. ■

We can now define the field $\mathbb{F}(X)$ of rational functions.

12.2.3 Remark. Let $(R, ')$ be a differentiable ring. The polynomial ring $R[X]$ is a differentiable ring if we assign X' arbitrarily and define

$$(X^n)' = nX^{n-1}X'$$

and extend linearly. If $(R, ')$ is the differentiable field $(\mathbb{F}, ')$, then the field $\mathbb{F}(X)$ of rational functions can be made into a differentiable field similarly if we define

$$\left(\frac{p(X)}{q(X)}\right)' = \frac{q(X)(p(X))' - p(X)(q(X))'}{(q(X))^2}.$$ ■

12.2.4 Example. Let R be a ring. Then $(R, ')$ is a differentiable ring, called the *trivial differentiable ring*, if for all $f \in R$, $f' = 0$. ■

12.2.5 Example. The set of infinitely differentiable functions from $\mathbb{R}$ to $\mathbb{R}$ is a differentiable ring if $'$ is the usual $\frac{d}{dx}$. ■

Now we are in a position to introduce some elementary functions such as the exponential and the logarithm.

12.2.6 Definition. Let $(\mathbb{F},')$ be a differentiable field and $f \in \mathbb{F}$. An element $y \in \mathbb{F}$ is said to be an *exponential of* f, denoted by $\exp(f)$, if $y' = f'y$. ■

Once we have an exponential, we also have *hyperbolic functions*. Recall that

(i) $\sinh(X) = \dfrac{\exp(X) - \exp(-X)}{2}$;

(ii) $\cosh(X) = \dfrac{\exp(X) + \exp(-X)}{2}$;

(iii) $\tanh(X) = \dfrac{\sinh(X)}{\cosh(X)}$.

We shall assume that the field $\mathbb{C}$ is a subfield of $\mathbb{F}$.

Now similarly we have *trigonometric functions* by recalling that

(iv) $\sin(X) = \dfrac{\exp(iX) - \exp(-iX)}{2i}$;

(v) $\cos(X) = \dfrac{\exp(iX) + \exp(-iX)}{2}$;

(vi) $\tan(X) = \dfrac{\sin(X)}{\cos(X)}$.

In an obvious way we can define the *inverse trigonometric functions* arcsin, arccos, arctan and the *inverse hyperbolic functions* arcsinh, arccosh, and arctanh.

12.2.7 Definition. Let $(\mathbb{F},')$ be a differentiable field and $f \in \mathbb{F}$. An element $y \in \mathbb{F}$ is said to be a *logarithm of* f, denoted by $\log(f)$, if $y' = \dfrac{f'}{f}$. ■

Finally, in a natural way, we can introduce, an integral on a differentiable field.

12.2.8 Definition. Let $(\mathbb{F},')$ be a differentiable field and $f \in \mathbb{F}$. An element $y \in \mathbb{F}$ is said to be an *integral of* f, denoted by $\int f$, if $y' = f$. ■

12.2.9 Definition. If $(\mathbb{E},')$ and $(F,')$ are differentiable fields such that $\mathbb{F}$ is a subfield of $\mathbb{E}$. and the derivation on $\mathbb{F}$ agrees with the derivation on $\mathbb{E}$ for $f \in \mathbb{F}$, then $(\mathbb{E},')$ is said to be a *differentiable extension field* of $\mathbb{F}$. ■

12.2.10 Remark. It is helpful to consider the following example. Let us start with the field $\mathbb{C}$. Then $\mathbb{C}[X]$ is the set of all polynomials with coefficients in $\mathbb{C}$, and is a ring. Indeed with the usual differentiation $'$, $(\mathbb{C}[X],')$ is a differentiable ring. Then $\mathbb{C}(X)$ is the set of all rational functions over $\mathbb{C}$ and with the usual differentiation $'$ is the differentiable field $(\mathbb{C}(X),')$. Observe that $\mathbb{C}(X) \supset \mathbb{C}[X]$. So we see that the usual derivation on $\mathbb{C}$ can be extended to $\mathbb{C}(X)$. $(\mathbb{C}(X),')$ is a differentiable extension field of $(\mathbb{C},')$.This is a special case of Proposition 12.2.12 (which we shall not prove). ■

12.2.11 Definition. A field $\mathbb{F}$ with multiplicative identity 1 and additive identity 0 is said to have *characteristic* 0 if $n \cdot 1 \neq 0$ for all $n \in \mathbb{N}$. ■

The fields $\mathbb{Q}$, $\mathbb{R}$, and $\mathbb{C}$ are fields of characteristic 0.

12.2.12 Proposition. *Let $(\mathbb{F},')$ be a differentiable field and t algebraic over the field $\mathbb{F}$. If the field $\mathbb{F}$ has characteristic 0, then the derivation $'$ on $\mathbb{F}$ extends to a derivation on $\mathbb{F}(t)$, which is also denoted by $'$, such that $(\mathbb{F}(t),')$ is a differentiable extension field of $(\mathbb{F},')$.* ■

12.2.13 Definition. Let $(\mathbb{F},')$ be a differentiable field of characteristic zero. A differentiable field $(\mathbb{E},')$ is said to be an *elementary differentiable extension field* of $(\mathbb{F},')$ if $\mathbb{E} = F(t_1, t_2, \ldots, t_n)$ where each t_i is algebraic over $\mathbb{F}(t_1, t_2, \ldots, t_{i-1})$ or is an exponential or a logarithm. If at least one of the t_i is not an exponential or a logarithm or algebraic over $\mathbb{F}(t_1, t_2, \ldots, t_{i-1})$, then $(\mathbb{E},')$ is said to be a *transcendental elementary extension field of* $(\mathbb{F},')$. ■

We are particularly interested in the case that $\mathbb{F} = \mathbb{C}(X)$ with the usual differentiation as derivation. The central question then is that if f is a member of some elementary differentiable extension field of $(\mathbb{C}(X),')$ is an integral $\int f$ a member of some (probably larger) elementary differentiable extension of $(\mathbb{C}(X),')$?
This is often expressed as: *when is an integral of an elementary function an elementary function? And if it is, can we calculate that integral?*

The American mathematician Maxwell Alexander Rosenlicht (1924–1999) in Rosenlicht (1972) used this algebraic formulation to prove the following result. (For a proof see Geddes et al. (1992).)

12.2.14 Theorem. *Let $(\mathbb{F},')$ be a differentiable field of characteristic zero and let $f \in \mathbb{F}$. If the equation $g' = f$ has a solution in some elementary differentiable extension field $(\mathbb{E},')$ of $(\mathbb{F},')$ where $(\mathbb{E},')$ and $(\mathbb{F},')$ have the same subfield C of constants, then there are constants $c_1, c_2, \ldots, c_n \in \mathbb{F}$ and elements $u_1, u_2, \ldots, u_n, v \in \mathbb{F}$ such that $f = \sum\limits_{i=1}^{n} c_i(u_i')/(u_i)v'$. In such a case,*

$$\int f = v + \sum_{i=1}^{n} c_i \log u_i$$

■

As a consequence of this theorem he showed that the integral we mentioned at the beginning of our discussion, namely $\int e^{x^2}\,dx$ and the following integrals are not elementary: $\displaystyle\int \frac{e^z}{z}\,dz$, $\displaystyle\int \frac{1}{\log z}\,dz$, and $\displaystyle\int \frac{\sin z}{z}\,dz$.

12.2.15 Remarks. A giant leap forward, in what is referred to as *computer algebra*, and in particular *symbolic integration*, was taken by the American mathematician Robert Henry Risch (born 1939) (whose PhD Advisor had been Rosenlicht) in a series of papers during the period 1968–1979. In over 100 pages he described an integration algorithm which he called a "*decision procedure*" which greatly extended the class of integrals which one could analyze and, if a solution exists, find a solution.

By the end of the twentieth century there were sophisticated general purpose computer algebra packages which implemented at least a significant portion of Risch decision procedure. These included Mathematica, Reduce, Maple, and Macsyma on microcomputers and mainframes There are extensions to Risch's algorithm, in particular the Risch-Norman algorithm which is less powerful but simpler and faster in that it avoids some recursion. It was discovered by the British computer scientist Arthur C. Norman in 1976.

According to WolframAlpha, even today Risch's decision procedure is not implemented fully in any computer package or app.

The powerful and very easy to use WolframAlpha app is available free of cost or at a nominal price for use on smartphones, tablets, laptops, and desktop computers. If you type in an integral, it returns not only an answer if it can, but also the intermediate steps it took to arrive at that answer.

Exercises 12.2

1. If $+, \cdot$ are the field operations on a field $\mathbb{F}$, and $(\mathbb{F}, ')$ is a differentiable field, then the set of constants, const$(\mathbb{F}, ')$, of $(\mathbb{F}, ')$ is a subfield of $\mathbb{F}$ (with the field operations $+, \cdot$.
2. Verify that if $\mathbb{F}$ is a subfield of $\mathbb{C}$, then the set of rational functions over $\mathbb{F}$ is a field.
3. If $(R, ')$ is a differentiable ring, where R is the ring $\mathbb{Z}$ of all integers or the field $\mathbb{Q}$ of all rational numbers, prove that $'$ is the trivial derivation. [Hint. Observe that $1 \cdot 1 = 1$ and consider $1'$.]

Additional Reading for Chapter 12

The constructibility of regular polygons is studied in (Clark, 1971, Sections 135–138), (Fraleigh, 1982, Section 48) and (Shapiro, 1975, Section 10.2).

Borwein and Crandall (2013) is a very readable and informative article on the term closed form.

The books Kaplansky (1957) and Knuth (1997) provide an introduction to differentiable fields by superb expositors. The article by A.C. Norman in Buchberger et al. (1983) gives an account of algorithms for integration in closed form, suitable for implementation on a computer, while the book Davenport (1981) presents the underlying mathematical theory. The paper Kasper (1980) has a very readable discussion of Integration in Finite Terms up to 1980, in particular covering the work of Joseph Liouville and touching on subsequent contributions by the Russian mathematician Alexander Markowich Ostrowski (1893–1986), Joseph Fels Ritt, Maxwell Alexander Rosenlicht, and Robert Henry Risch. The books Bronstein (1997); Geddes et al. (1992) and the paper Cherry (1985) together provide a wide coverage of computer algebra.

References

Abel, N.: Mémoire sur une classe particulière d'équations résolubles algébriquement. J. reine und angewandte Mathematik **4**, 131–156 (1826)
Adamchik, V.S., Jeffrey, D.J.: Polynomial transformations of Tschirnhaus, Bring and Jerrard. ACM SIGSAM Bull. **37**, 90–94. https://tinyurl.com/ybazbzx3 (2003)
Adamson, I.: Introduction to Field Theory, 4th edn. Dover, New York (2007)
Aigner, M., Ziegler, G.: Proofs from THE BOOK, with Illustrations by Karl H. Springer, Hofmann (2018)
Archbold, J.: Algebra, 2nd edn. Pitman, London (1970)
Artin, E., Milgram, A.: Galois Theory, 2nd edn. University of Notre Dame Press, London, Notre Dame (1971)
Baker, A.: Transcendental Number Theory. Cambridge University Press, Cambridge (1975)
Beardon, A., Ng, T.: On Ritt's factorization of polynomials. J. Lond. Math. Soc. **62**, 127–138 (2000). https://doi.org/10.1112/S0024610700001046
Beckmann, P.: A History of π. St. Martin's Press, New York (1971)
Bell, E.: Men of Mathematics. Simon and Schuster, New York (1937)
Bell, E.: The Development of Mathematics. McGraw Hill, New York (1945)
Berggren, L., Borwein, J., Borwein, P.: Pi: A Source Book. Springer, New York (1997)
Beukers, F., Bezivia, J., Robba, P.: An alternative proof of the Lindemann-Weierstrass theorem. Am. Math. Mon. **97**, 193–197 (1990)
Bewersdorff, J.: Galois Theory for Beginners. Translated by David Kramer. Am. Math. Soc, Providence (2006)
Birkhoff, G., MacLane, S.: A Survey of Modern Algebra. Macmillan, New York (1953)
Blum-Smith, B., Coskey, S.: The fundamental theorem on symmetric polynomials: History's first Whiff of Galois Theory. 16pp (2016). arXiv:1301.7116v4
Bold, B.: Famous Problems of Geometry and How to Solve Them. Dover, New York (1969)
Borwein, J., Crandall, R.E.: Closed forms: what they are and why we care. Notices Am. Math. Soc. **60**, 50–65 (2013)
Boyer, C.: A History of Mathematics. Wiley, New York (1968)
Bronstein, M.: Symbolic Integration I: Transcendental Functions. Springer, New York (1997)
Brooks, M.: The Quantum Astrologer's Handbook. Scribe Publications, Brunswick (2017)
Browder, F.E. (ed.): Mathematical Developments Arising from Hilbert Problems, Part 1. Am. Math. Soc, Rhode Island (1976)
Browder, F.E. (ed.): Mathematical Developments Arising from Hilbert Problems, Part 2. Am. Math. Soc, Rhode Island (1976)

S. A. Morris et al., *Abstract Algebra and Famous Impossibilities*, Undergraduate Texts in Mathematics, https://doi.org/10.1007/978-3-031-05698-7

Buchberger, B., Collins, G., Loos, R., Albrecht, R. (eds.): Computer Algebra: Symbolic Algebraic Computation, 2nd edn. Springer, Vienna (1983)
Cardano, G.: Ars Magna or the Rules of Algebra. Translated from Latin into English and edited by T. Richard Witmer of the 1565 edition with additions from the 1570 and 1663 editions. Dover Publications Inc., New York (1993)
Chalebgwa, T.P., Morris, S.A.: Topological transcendental fields. Axioms **11**, 118 (2022). https://doi.org/10.3390/axioms11030118
Chalebgwa, T.P., Morris, S.A.: n.d., Sin, cos, exp, and log of Liouville numbers. arXiv:2202.11293 (to appear)
Cherry, G.: Integration in finite terms with special functions: the error functions. J. Symb. Comput. **1**, 283–302 (1985)
Clark, A.: Elements of Abstract Algebra. Wadsworth, Belmont (1971)
Courant, R., Robbins, H.: What is Mathematics? Oxford University Press, New York (1941)
Cox, D.: Galois Theory, 2nd edn. Wiley, Hoboken (2012)
Cox, D.A.: Why Eisenstein proved Eisenstein's criterion and why Schönemann discovered it first. Am. Math. Mon. **118**, 3–21 (2011)
Davenport, J.: On the integration of algebraic functions. In: Lecture Notes in Computer Science, vol. 102. Springer, Berlin (1981)
De Morgan, A.: A Budget of Paradoxes. Dover, New York (1954)
Dubins, L., Hirsch, M., Karush, J.: Scissor congruence. Isr. J. Math. **1**, 239–247 (1963)
Dudley, U.: A Budget of Trisections. Springer, New York (1987)
Dörrie, H.: 100 Great Problems of Elementary Mathematics. Dover, Translated by David Antin (1965)
Dummit, D.: Solving solvable quintics. Math. Comput. **57**, 387–401 (1991)
Edwards, H.: Galois Theory. Springer, Heidelberg (1984)
Edwards, H.: Roots of solvable polynomials of prime degree. Expo. Math. **32**, 79–91 (2014)
Eisenstein, F.: Über die Irreductibilität und einige andere Eigenschaften der Gleichung, von welcher die Theilung der ganzen Lemniscate abhängt. J. reine und angewandte Math. **39**, 160–179 (1850)
Eves, H.: A Survey of Geometry. Allyn and Bacon, Boston (1972)
Ferrar, W.: Higher Algebra. Clarendon, Oxford (1958)
Fine, B., Rosenberger, G.: The Fundamental Theorem of Algebra. Springer, New York (1997)
Fraleigh, J.: A First Course in Abstract Algebra, 3rd edn. Addison-Wesley, Reading (1982)
French, R.: The Banach-Tarski theorem. Math. Intell. **10**, 21–28 (1988)
Gallian, J.: Contemporary Abstract Algebra, 9th edn. Cengage Learning, Australia (2017)
Garcia, S.R., Miller, S.J.: 100 Years of Math Milestones: The Pi Mu Epsilon Centennial Collection. Am. Math. Soc, Providence (2019)
Gardner, M.: Mathematical Circus. Penguin Books, Middlesex (1981)
Gauss, C.F.: Latin version appeared in 1801, Disquisitiones Arithmeticae. Translated from Latin into English by Arthur A. Clarke. Yale University Press, New Haven (1966)
Geddes, K.O., Czapor, S.R., Labahn, G.: Algorithms for Computer Algebra. Kluwer Academic Publishers, Boston (1992)
Gelfond, A.: Transcendental and Algebraic Numbers (translated by Leo F. Boron). Dover Publications (2015)
Gilbert, W.: Modern Algebra with Applications. Wiley, New York (1976)
Gordan, P.: Transcendenz von e und π. Math. Ann. **43**, 222–224 (1893)
Gouvêa, F.Q.: A Guide to Groups, Rings, and Fields. Mathematical Association of America, Washington, DC (2012)
Gow, J.: A Short History of Greek Mathematics. Chelsea, New York (1968)
Guilbeau, L.: The history of the solution of the cubic equation. Math. News Lett. **5**, 8–12 (1930)
Hadlock, C.: Field Theory and Its Classical Problems, Carus Mathematical Monographs, No. 19. Mathematical Association of America (1978)
Halmos, P.: Naive Set Theory. Van Nostrand Reinhold Company, New York, Cincinnati (1960)

Hamilton, W.R.: Inquiry into the validity of a method proposed by George B. Jerrard, Esq., for transforming and resolving equations of elevated degrees. British Association for the Advancement of Science August 1836 report, Bristol, pp. 295–348. https://www.maths.tcd.ie/pub/HistMath/People/Hamilton/Jerrard/Jerrard.pdf (1836)
Hardy, G.: A Course of Pure Mathematics, 10th edn. Cambridge University Press, Cambridge (1952)
Hardy, G., Wright, E.: An Introduction to the Theory of Numbers. Clarendon, Oxford (1960)
Harley, R.R.: A contribution to the history of the problem of the reduction of the general equation of the fifth degree to a trinomial form (with an Appendix on Bring's reduction of the quintic equation). Q. J. Pure Appl. Math. **6**, 38–47 (1864). tinyurl.com/Harley1864
Heath, T.: A History of Greek Mathematics Vol. I & II. Clarendon Press, Oxford (1921)
Hermite, C.: Sur la fonction exponentielle. Comptes Rendus des Séances de l'Académie des Sciences Paris **77**, 18–24 (1873)
Herstein, I.: Topics in Algebra. Blaisdell, New York (1964)
Hilbert, D.: Über die transcendenz der zahlen e und π. Math. Ann. **43**, 216–219 (1893); reprinted in Gesammelte Abhandlungen, vol.1. Chelsea (1965)
Hobson, E.: Squaring the Circle. Cambridge University Press (1913) reprinted in Squaring the Circle, and Other Monographs. Chelsea (1953)
Holden, H.: Rings on $\mathbb{R}^2$. Math. Mag. **62**, 48–51 (1989)
Hudson, H.: Ruler and Compass. Longmans Green (1916) reprinted in Squaring the Circle, and Other Monographs. Chelsea (1953)
Hurwitz, A.: Beweis der transcendenz der zahl e. Math. Ann. **43**, 220–222 (1893)
Jacobs, H.: Geometry. Freeman, New York (1953)
Jagy, W.: Squaring circles in the hyperbolic plane'. Math. Intell. **17**, 31–36 (1995)
Jerrard, G.: An Essay on the Resolution of Equations. Taylor and Francis, London. https://ia800306.us.archive.org/18/items/essayonresolutio00jerrrich/essayonresolutio00jerrrich.pdf (1859)
Jones, W.: First published 1706. Synopsis Palmariorum Matheseos or New Introduction to Mathematics, Gale ECCO (2018)
Kaplansky, I.: An Introduction to Differential Algebra. Hermann, Paris (1957)
Kasper, T.: Integration in finite terms: the Liouville theory. Math. Mag. **53**, 195–201 (1980)
Klein, F.: Elementary Mathematics from an Advanced Standpoint, (vol 1: Arithmetic, Algebra and Analysis). Dover, New York (1948)
Klein, F.: Famous Problems of Elementary Geometry; reprinted in Famous Problems, and Other Monographs. Chelsea, San Francisco (1962)
Kline, M.: Mathematical Thought from Ancient to Modern Times. Oxford University Press (1972)
Knuth, D.E.: The Art of Computer Programming: Volume 2 Seminumerical Algorithms, 3rd edn. Addison-Wesley Publishing Company, Reading (1997)
Kostovskii, A.: Geometrical Constructions Using Compasses Only. Blaisdel (1961)
Laczkovich, M.: Equidecomposability and discrepancy: a solution of tarski's circle-squaring problem. J. reine und angewandte Mathematik **404**, 77–117 (1990)
Lang, S.: Introduction to Transcendental Numbers. Addison-Wesley (1966)
Lang, S.: Algebra, 3rd edn. Springer, New York (2002)
Lasserre, F.: The Birth of Mathematics in the Age of Plato. Hutchinson, London (1964)
Ledermann, W.: Complex Numbers. Library of Mathematics. Routledge and Kegan Paul, London (1965)
Lindemann, F.: Über die zahl π. Math. Ann. **20**, 213–225 (1882)
Lorenz, F.: Algebra, Volume 1: Fields and Galois Theory. Springer Science, USA (2006)
Macdonald, I.: Symmetric Functions and Hall Polynomials, 2nd edn. Oxford University Press, Oxford (2015)
MacLane, S., Birkhoff, G.: Algebra. Chelsea Publishing Company, New York (1988)
Mahler, K.: Lectures on Transcendental Numbers, Springer Lecture Notes in Mathematics 546. Springer, New York (1976)
Marks, A.S., Unger, S.T.: Borel circle squaring. Ann. Math. **186**, 581–605 (2017)

Mascheroni, L.: La geometria del compasso. https://babel.hathitrust.org/cgi/pt?id=osu.32435018455220&view=1up&seq=7 (1797)
Morris, S.: Topology without tears. http://www.topologywithouttears.net (2020)
Murty, M.R., Rath, P.: Transcendental Numbers. Springer Nature, Switzerland (2014)
Natarajan, S., Thangadurai, R.: Pillars of Transcendental Number Theory. Springer, New York (2020)
Neumann, P.M.: The Mathematical Writings of Évariste Galois. European Mathematical Society, Zürich (2010)
Newton, S.I.: Universal arithmetick: or, A treatise of arithmetical composition and resolution an English translation of Arithmetic Universalis published in Latin in 1707. London (1769)
Niven, I.: The transcendence of π. Am. Math. Mon. **46**, 469–471 (1939)
Niven, I.: Numbers: Rational and Irrational. Random House, New York (1961)
Niven, I.: Irrational Numbers, Carus Mathematical Monographs, No. 11. Mathematical Association of America (1967)
Pan, Y., Chen, Y.: On Kronecker's Solvability Theorem (2020). arXiv:1912.07489v6
Parshin, A.: Number Theory IV: Transcendental Numbers (Encyclopaedia of Mathematical Sciences (44)). Springer, New York (1997)
Pearce, P., Ramsay, J., Roberts, H., Tinoza, N., Willert, J., Wu, W.: The Circle Squaring Problem Decomposed. Mathematical Horizons p, November issue (2009)
Prasolov, V.V.: Polynomials. Springer, New York (2004)
Ribenboim, P.: My Numbers, My Friends: Popular Lectures on Number Theory. Springer, New York (2000)
Risch, R.: The solution of the problem of integration in finite terms. Trans. Am. Math. Soc. **76**, 605–608 (1970)
Rosen, M.I.: Niels Hendrik Abel and equations of the fifth degree. Am. Math. Mon. **102**, 495–505 (1995)
Rosenlicht, M.: Integration in finite terms. Am. Math. Mon. **79**, 963–972 (1972)
Rothman, T.: Genius and biographers: the fictionalization of Évariste Galois. Am. Math. Mon. **89**, 84–106 (1982)
Rotman, J.: Galois Theory, 2nd edn. Springer, New York (2001)
Ruthen, R.: Squaring the circle. Sci. Am. **261**, 11–12 (1989)
Sanford, V.: A Short History of Mathematics. Harrap, London (1958)
Schönemann, T.: Von denjenigen Moduln, welche Potenzen von Primzahlen sind. J. reine und angewdte Math. **32**, 93–105 (1846)
Sethuraman, B.: Rings, Fields and Vector Spaces: An Introduction to Abstract Algebra Via Geometric Constructibility. Springer, New York (1997)
Shapiro, L.: Introduction to Abstract Algebra. McGraw-Hill, New York (1975)
Shipman, J.: Improving the fundamental theorem of algebra. Math. Intell. **29**, 9–14 (2007)
Sigler, L.: Fibonacci's Liber Abaci: A Translation into Modern English of Leonardo Pisano's Book of Calculation. Springer Nature, USA (2003)
Smith, D.: The History and Transcendence of π; reprinted in W.A. Young, Monographs on Topics of Modern Mathematics Relevant to the Elementary Field. Dover, New York (1955)
Smoryński, C.: History of Mathematics: A Supplement. Springer, New York (2008)
Spearman, B.K., Williams, K.S.: Characterization of solvable quintics $x^5 + ax + b$. Am. Math. Mon. **101**, 986–992 (1994)
Spivak, M.: Calculus. W.A. Benjamin, New York (1967)
Steiner, J.: Geometrical constructions with a ruler given a fixed circle with its center. Scripta Math. **14**, 187–264 (1948)
Stewart, I.: Galois Theory, 3rd edn. Chapman & Hall, Boca Raton (2004)
Stewart, I.: Visions of Infinity: The Great Mathematical Problems. Basic Books, New York (2013)
Stewart, I.: Significant: Lives and Works of Trailblazing Mathematicians. Profile Books, London (2017)
Struik, D.: A Concise History of Mathematics. Dover, New York (1967)

Sturm, J.: Mémoire sur la résolution des équations numériques'. Bull. des Sci. de Férussac **11**, 419–425 (1829)
Tignol, J.-P.: Galois' Theory of Algebraic Equations. World Scientific, Singapore (2002)
Tropper, A.: Linear Algebra. Thomas Nelson, London (1969)
Turnbull, H.: The Great Mathematicians. Methuen, London (1933)
von Tschirnhaus, E.W.: A method for removing all intermediate terms from a given equation. Acta Erud. 204–207 (1683). Translated into English by R.F. Green and republished in ACM SIGSAM Bulletin (37), 1–3 (2003)
Wagon, S.: Circle-squaring in the twentieth century. Math. Intell. **3**, 176–181 (1981)
Wantzel, M.: Recherches sur les moyens de reconnaître si un problème de géométrie peut se résoudre avec la. règle et le compas. J. de Mathématiques Pures et Appliquées **2**, 372–466 (1837)
Weierstrass, K.: Zu lindemann's abhandlung. 'Über die ludolph'sche zahl'. Sitzungsberichte der Königlich Preussischen Akademie der Wissen-schaften zu Berlin **5**, 1067–1085 (1885)
Wilder, R.L.: Introduction to the Foundations of Mathematics, 2nd edn. Dover (2012)
Young, R.: When is $\mathbb{R}^n$ a field? Math. Gaz. **72**, 128–129 (1988)

Index

Page references in bold indicate locations where formal statements are given.

S. A. Morris et al., *Abstract Algebra and Famous Impossibilities*, Undergraduate Texts in Mathematics, https://doi.org/10.1007/978-3-031-05698-7

Printed in the United States
by Baker & Taylor Publisher Services